Barriers and Facilitators of Geriatric Smart Home Technology Implementation

Barriers and Facilitators of Geriatric Smart Home Technology Implementation

"Breaking Down the Walls: Overcoming Obstacles to Smart Home Technology for Seniors"

S. K. MEENA

NEHA JAIN

YASH JAIN

MOTILAL BANARSIDASS INTERNATIONAL DELHI

First Edition: Delhi, 2024

© MOTILAL BANARSIDASS INTERNATIONAL
All Rights Reserved

ISBN : 978-81-19394-62-3

Also available at
MOTILAL BANARSIDASS INTERNATIONAL
H. O. : 41 U.A. Bungalow Road, (Back Lane)Jawahar Nagar, Delhi - 110 007
4261 (basement) Lane #3,Ansari Road, Darya Ganj, New Delhi - 110 002
203 Royapettah High Road, Mylapore, Chennai - 600 004
12/1A, 2nd Floor, Bankim Chatterjee Street, Kolkata - 700 073
Stockist : Motilal Books, Ashok Rajpath, Near Kali Mandir, Patna - 800 004

No part of this book may be reproduced in any form or by any electronic or mechanical means including information storage and retrieval systems without permission in writing from the publishers, excepts by a reviewer who may quote brief passages in a review.

Printed in India
MOTILAL BANARSIDASS INTERNATIONAL

Dedication

In Loving Memory of

Shrimati Soma Devi Meena

Inspiration in Absence

This book is lovingly dedicated to my late mother, Shrimati Soma Devi Meena, whose love, knowledge, and strength continue to inspire me daily. Though physically absent, her spirit remains an unwavering guide through the pages of this book. In her memory, and with gratitude for the profound impact she had on my life, I honor her legacy. Though greatly missed, she will forever be cherished and loved.

S.K. Meena

In Honor of the Seniors

This book is dedicated to all the seniors who have paved the way for the generations that followed. Your wisdom, experience, and resilience have been a beacon of inspiration for us all. May this book stand as a tribute to your enduring contributions and a valuable resource aimed at enhancing the quality of life for seniors. In celebrating your journey, we aspire to contribute to a future where technology and compassion intertwine to create a more vibrant and empowered aging experience.

With heartfelt dedication,

S.K. Meena
Neha Jain
Yash Jain

Foreword

This book, "Barriers and Facilitators of Geriatric Smart Home Technology Implementation," delves into the intricate interplay between technological barriers, sociocultural factors, and various challenges that older individuals encounter in embracing smart home technologies. Through extensive research and thoughtful analysis, it identifies key facilitators crucial in aiding seniors to effectively adopt and utilize these technologies.

As we embark on this journey together, let's explore the nuances of geriatric smart home technology implementations, aiming to dismantle obstacles and leverage opportunities. The vision is to shape a future where individuals of all ages can coexist comfortably and confidently with technology.

This handbook serves as a beacon for the elderly population, a testament to the unwavering dedication of medical professionals, engineering technologists, physical therapists, and occupational therapists. Their tireless efforts aim to enhance the quality of life for communities and individual families alike.

"Barriers and Facilitators of Geriatric Smart Home Technology Implementation" sheds light on the function and significance of smart home technology in enriching the independence and quality of life for the elderly. It is a call to explore how the remarkable technological strides of the past decade can be harnessed to uplift the well-being of our aging population.

I extend my commendation to Prof. S. K. Meena, Dr. Neha Jain, and Er. Yash Jain for their unwavering commitment to advancing knowledge in this crucial area. Their efforts to improve the quality of life for older individuals through the thoughtful implementation of smart home technologies are truly commendable.

With warm regards,

Dr. Ganesh Narain Saxena
Emeritus Professor of General Medicine
Former Pro - Vice-Chancellor
MGUMST, Jaipur

Preface

Welcome to the unfolding exploration of a domain gaining increasing relevance: the integration of smart home technologies into the living environments of the elderly. Within the pages of this book, we delve into the multifaceted aspects of adopting such technologies, shedding light on the barriers and facilitators that shape how seniors incorporate these innovations into their daily lives.

The imperative for inventive solutions to foster independent living and enhance the quality of life for our aging society has never been more pressing. Smart home technologies hold tremendous promise in this regard, yet their effective implementation is often impeded by various obstacles. This book undertakes a comprehensive examination of the intricate interplay between technological barriers, socio-cultural factors, and other challenges through extensive research and analysis.

Beyond unveiling the hindrances, this work strives to identify key facilitators that can empower older individuals to embrace and utilize smart home technologies effectively. Together, let us embark on a journey to explore the nuanced landscape of geriatric smart home technology implementations. Our collective endeavor is to dismantle barriers and seize opportunities, contributing to the creation of a future where individuals of all ages can live comfortably and confidently with technology.

This book serves as a guide, inviting you to join us on this transformative journey, as we navigate the complexities, uncover solutions, and envision a world where the integration of smart home technologies becomes a seamless and empowering aspect of aging gracefully.

Acknowledgments

Writing a book is a profound journey, and our gratitude extends to those whose unwavering support has made "Barriers and Facilitators of Geriatric Smart Home Technology Implementation" (Breaking Down the Walls: Overcoming Obstacles to Smart Home Technology for Seniors) possible.

Firstly, we express our heartfelt thanks to our families for their enduring love, support, and patience. Their encouragement, understanding, and sacrifices have been the bedrock of this endeavor.

We extend our appreciation to Mahatma Gandhi University of Medical Sciences & Technology, Jaipur, for providing resources, facilities, and invaluable opportunities that have significantly contributed to the content of this book.

Our sincere thanks to the esteemed leadership of the University—Hon'ble Emeritus Chairperson and founder, Dr. M L Swarankar, respected Chairperson cum Chancellor, Dr. Vikas Chandra Swarankar, President cum Vice-Chancellor, Dr. Achal Gulati, Pro-Vice-Chancellor, Dr. V K Kapoor, and Registrar, Dr. A K Sharma.

Special mention goes to former Vice-Chancellor Dr. Sudhir Sachdev, former Pro-Vice-Chancellor, Dr. G N Saxena, and Dr. Munish Kumar Kakkar, Professor of Paediatrics at Mahatma Gandhi Medical College, Jaipur.

Our gratitude extends to Dr. Pankaj Bajpai, Ex HOD at National Institute for Locomotor Disabilities, Kolkata, and President of the All India Occupational Therapists' Association, for his mentorship and guidance.

We are indebted to our editors and reviewers for their critical feedback, insightful suggestions, and constructive criticism, **refining our ideas and elevating the overall quality** of the book.

A heartfelt thank you to the All India Occupational Therapists' Association for their continuous support and encouragement.

Lastly, to our readers - your keen interest and engagement make our efforts worthwhile. Thank you for giving us the opportunity to share our knowledge and experiences with you.

To everyone who contributed and supported us on this journey, thank you for being an integral part of this endeavor.

With gratitude,

S.K. Meena
Neha Jain
Yash Jain

About the Authors

The authors of the inaugural book, "Barriers and Facilitators of Geriatric Smart Home Technology Implementation," are a dynamic trio of experts blending backgrounds in occupational therapy and engineering, united by a shared passion for enhancing the lives of seniors through technological advancements.

Prof. S K Meena—A distinguished Professor and Principal, Mahatma Gandhi Occupational Therapy College, Director Mahatma Gandhi Institute of Allied Health Sciences—a unit of Mahatma Gandhi University of Medical Sciences and Technology, Jaipur, and also Director SNM Autism Research Institute and SNM Institute of Brain & Spine Rehabilitation, Jaipur. He is also an executive committee member of the Society of Hand Therapy India and has served as a resource person for National or International Conferences, Seminars, and Workshops. Prof. Meena has received 32 distinguished awards and honours for his outstanding contribution to research in occupational therapy. He has published 30 research papers in national and international journals. Prof. Meena is also a finance and business expert and holds other memberships in different categories in Special Olympics Bharat. Prof. Meena has extensive experience in the field of occupational therapy.

Dr. Neha Jain—An Associate Professor at Mahatma Gandhi Occupational Therapy College, Dr. Jain is a trailblazer in her own right. Holding a PhD in Occupational Therapy, she stands as the first person from Rajasthan to achieve

this distinction. With over a decade of experience, Dr. Jain is an active member of the All India Occupational Therapists Association (AIOTA). Her contributions extend beyond academia, earning her numerous accolades for clinical and research excellence, including a humanitarian award. Driven by a commitment to societal well-being, she actively supports underprivileged communities.

Er. Yash Jain—A marine engineer officer in the Merchant Navy, Er. Yash Jain brings a unique perspective to the collaboration. Armed with a Bachelor's degree in Mechanical  Engineering and a Post Graduate Diploma in Marine Engineering, his keen eye for detail and strong work ethic have been invaluable in the research and writing process. As the co-author of the book, his technical skills add depth to the exploration of geriatric smart home technology.

Together, these authors amalgamate their diverse expertise to deliver a comprehensive understanding of geriatric smart home technology implementation. In their debut publication, they offer practical solutions and strategies, illuminating the path to overcome barriers faced by seniors and uncovering the **potential benefits that technology can provide in enriching the lives of the elderly.**

Contents

Dedication .... (v)

Foreword .... (vii)

Preface .... (ix)

Acknowledgments .... (xi)

About the Author .... (xiii)

Introduction 1

Ageing—The Old Age 3

Geriatrics and Quality of Life 5

Ageing and Technology 8

Smart Home Technology 9

Geriatric Smart Home Technology 12

Overview of the Importance of
Geriatric Smart Home Technology 14

Current State of Geriatric Care: Challenges
and Opportunities for Improvement 18

Understanding Geriatric Smart Home Technology:
Key Features and Benefits 22

Barriers to Geriatric Smart Home Technology
Implementation: Financial, Technological,
and Societal Challenges 31

Facilitators of Geriatric Smart Home Technology
Implementation: Strategies for Success 40

Case Studies: Real-World Examples
of Successful Implementation.. 49

Ethical Considerations: Balancing Autonomy
and Privacy with Technological Advancements 60

Future of Geriatric Smart Home Technology:
Innovations and Trends 64

Geriatric Smart Home Technology in India:
Current Status, Future Prospects, and Challenges 75

Reflections on the Impact of Geriatric Smart Home
Technology on Aging Populations 78

Conclusion 83

References 85

Introduction

"As our population continues to age, it is essential that we leverage the power of technology to enhance the quality of life for older adults. Geriatric smart home technology has the potential to transform the way we care for our seniors, improving their safety, health, and overall well-being."

Technology has revolutionized the way we live, work and communicate with one another. Smart home technology is one such advancement that has made life easier and more convenient for people of all ages. In particular, geriatric smart home technology has the potential to improve the quality of life for older adults, enabling them to live independently for longer and maintain their health and well-being.

However, the implementation of geriatric smart home technology is not without its challenges. Barriers such as cost, lack of awareness, and privacy concerns can hinder its adoption, while facilitators such as ease of use, accessibility, and customization can help overcome these barriers.

This book is a comprehensive guide to understanding the barriers and facilitators of geriatric smart home technology implementation. It provides practical insights and solutions for individuals, caregivers, healthcare professionals, and policymakers who are interested in utilizing technology to enhance the lives of older adults.

By addressing the challenges and opportunities of geriatric smart home technology, this book aims to provide readers with a deeper understanding of its potential benefits and limitations.

Through a series of case studies, expert insights, and practical tips, readers will learn how to overcome barriers to adoption and make the most of the technology to improve the lives of older adults.

As a working professional in the field, we have seen first-hand the positive impact that geriatric smart home technology can have on individuals and society as a whole. In this book, we share our insights and experiences to help readers navigate the complex landscape of geriatric smart home technology and make informed decisions about its implementation.

We encourage readers to approach this book with an open mind and a willingness to learn. Whether you are a healthcare professional, caregiver, or simply interested in improving the lives of older adults, this book offers practical solutions and insights that can make a real difference. So, don't wait any longer, let's dive into the world of geriatric smart home technology and discover how it can transform the lives of older adults.

Ageing—The Old Age

Ageing is a natural process that affects every individual as they grow older. It is marked by a gradual decline in physical, cognitive, and emotional functioning, and can be accompanied by a range of health issues and challenges. As individuals reach the later stages of life, they may experience what is commonly referred to as "old age." The term "ageing" encompasses a multidimensional process of change that affects a person's physical, psychological, and social spheres. This process is not a steady state and does not necessarily involve deterioration or decrement. Various definitions and terms are used interchangeably, leading to confusion and uncertainty regarding who qualifies as "elderly" or "aged" and at what age. Unfortunately, prevailing unfavorable attitudes towards ageing can exacerbate this confusion. Biological ageing refers to physical changes in the body that can begin long before an individual reaches the chronological age of 65. Psychological ageing involves an individual's adaptive capacities and self-awareness, while social ageing refers to social habits and roles within a specific culture or society. As social age progresses, individuals may experience a decline in meaningful social interaction. Affective ageing involves reduced adaptive capacities in relation to changes in the environment. The term "aged" is difficult to define, as researchers use different criteria to determine when an individual can be classified as such.

Old age is typically defined as the period of life after middle age, usually starting around age 65. While it is true that ageing can bring about new challenges and limitations, it is important to remember that many older adults continue to live fulfilling and meaningful lives.

One of the key challenges associated with old age is physical decline. As individuals age, they may experience a range of health issues such as arthritis, vision and hearing loss, and mobility problems. These issues can make it more difficult for older adults to engage in activities they once enjoyed, and may require them to rely on the help of others.

Cognitive decline is another common challenge associated with old age. As individuals age, they may experience changes in memory, attention, and decision-making abilities. While these changes can be frustrating, there are many ways to maintain cognitive function in later life, such as engaging in mental exercises and staying socially active.

Emotional health is also an important consideration in old age. Older adults may experience feelings of isolation, loneliness, and depression, particularly if they have lost loved ones or are no longer able to engage in activities they once enjoyed. It is important for older adults to stay connected with friends and family, and to seek help if they are experiencing mental health issues.

Summary

While old age can bring about new challenges and limitations, it is important to remember that many older adults continue to live fulfilling and meaningful lives. By staying physically active, mentally engaged, and socially connected, older adults can enjoy a happy and healthy later life.

Geriatrics and Quality of Life

The concept of life quality is based on an individual's personal feelings, experiences, outlook, and daily activities. This includes their level of happiness, comfort, security, productivity, and health. As professionals in aging, it's important to assess the quality of life by determining the personal meaning associated with different aspects of it. It's essential to understand that a positive outlook and quality of life in older adults doesn't necessarily indicate good mental or physical health. However, having a positive outlook and hope can lead to better outcomes.

Physical and mental impairments can significantly decrease the quality of life in older adults, causing loss of hope, depression, isolation, and lack of joy in everyday life. Care professionals should identify factors like depression, social isolation, or chronic medical illness and take immediate measures to improve the client's quality of life. This includes interventions to enhance their comfort level, happiness, social immersion, access to care, resources, and support systems.

Apart from formal quality of life assessment, it's crucial to measure it by identifying what the "ideal life" means to the clients. Engaging in conversations to discover their preferences, values, and things that fulfill them can help improve their quality of life. Verbal communication is a great way to assess quality of life and determine areas that need improvement. It's vital to individualize the approach to quality of life assessment based on personality, ability, and preference.

Geriatrics is a medical specialty that focuses on the health and well-being of older adults. Quality of life is an important aspect of geriatric care, as it is important for older adults to maintain a high level of physical, mental, and social functioning in order to enjoy their later years.

Geriatric care providers use a variety of strategies to improve the quality of life for older adults. These may include:

Medication management: Many older adults take multiple medications, which can increase the risk of adverse side effects and interactions. Geriatric care providers can help manage medications to ensure that they are safe and effective, while minimizing the risk of side effects.

Chronic disease management: Older adults are more likely to have chronic health conditions, such as diabetes, heart disease, and arthritis. Geriatric care providers can help manage these conditions through regular check-ups, medication management, and lifestyle changes.

Nutrition and exercise: A healthy diet and regular exercise can help older adults maintain physical and mental health, and reduce the risk of chronic disease. Geriatric care providers can provide guidance on nutrition and exercise that is tailored to the individual's needs and abilities.

Social support: Social isolation can have a negative impact on the quality of life for older adults. Geriatric care providers can help connect older adults with social support networks, such as senior centers, community groups, and support groups.

Cognitive and mental health: Geriatric care providers can also help manage cognitive and mental health issues, such as dementia, depression, and anxiety. This may include medication management, counseling, and cognitive training exercises.

Summary

Geriatric care providers use a variety of strategies to improve the quality of life for older adults. By managing medications, chronic conditions, nutrition and exercise, social support, and cognitive and mental health, geriatric care providers can help older adults maintain their physical, mental, and social functioning, and enjoy a high quality of life in their later years.

Ageing and Technology

What is the concept of "Aging in Place"? It refers to an individual's ability to continue living in their own home as they age, either independently or with the help of caregivers, assistive technology, or other support services. The current healthcare system is facing challenges, with the cost of healthcare increasing rapidly, making it necessary to find alternative, low-cost, and sustainable arrangements for healthcare. One solution to this issue is transforming regular homes into smart homes through ubiquitous computing technology, which can support the care of elderly individuals who live independently. Technological advancements have greatly enhanced the lives of the elderly, and these advancements are included in political strategies to address the challenges of an aging society. A study on the Evolution of Smart Homes for the Elderly categorizes smart home technologies into six areas: Health and Nutrition, Leisure, Personal Hygiene and Care, Physical Activity, Safety, and Social Engagement. One-third of these categories are aimed at downtime and social engagement, which is crucial as elderly individuals may become bored or isolated. Despite slower technology adoption among seniors, smart technology can help counter these issues. With the world's rapid population aging, there is a growing interest in "smart home" technologies that can assist older adults in continuing to live at home safely and independently. A systematic review and critical evaluation of worldwide literature assesses the effectiveness and feasibility of smart-home technologies in promoting independence, health, well-being, and quality of life among older adults.

Smart Home Technology

"Smart Home technology started for more than a decade to introduce the concept of networking devices and equipment in the house. According to the Smart Homes Association the best definition of smart home technology is: the integration of technology and services through home networking for a better quality of living. Many tools that are used in computer systems can also be integrated in Smart Home Systems. In this book, we present the battles and supported Technologies of Geriatric smart homes."

Smart home technology refers to the use of devices, systems, and appliances that are connected to a home network and can be controlled or automated through a mobile device, computer, or voice command. These devices are designed to make daily living more convenient, efficient, and secure by allowing users to remotely control and monitor their home environment.

Smart home technology typically includes a range of devices such as smart thermostats, smart lighting, smart locks, smart security cameras, smart appliances, and voice-activated assistants like Amazon's Alexa or Google Assistant. These devices are often interconnected through a central hub or app that allows users to control them from a single location.

For example, a smart thermostat can be programmed to adjust the temperature of a home based on the time of day, the occupancy of the home, and the user's preferences. A smart lighting system can be programmed to turn on or off automatically when someone enters or leaves a room. A smart lock can be controlled remotely, allowing homeowners to lock or unlock their doors from their phone or computer.

Smart home technology can also include sensors that detect changes in the home environment and send alerts to the homeowner's phone or computer. For example, a water leak sensor can detect a leak and send an alert to the homeowner's phone, allowing them to take action quickly to prevent water damage.

Smart home technology has had a significant impact on individuals, society, country, and the economy as a whole. Some of the key impacts of smart home technology include:

Improved quality of life: Smart home technology has made daily living more convenient, efficient, and comfortable for individuals. The ability to control and monitor home devices remotely has allowed individuals to better manage their time, reduce stress, and increase productivity. For older adults or people with disabilities, smart home technology can also provide greater independence and a higher quality of life.

Increased safety and security: Smart home technology has also improved safety and security in homes. Smart security systems can detect and alert homeowners to potential security threats, while smart smoke detectors can detect smoke and carbon monoxide and send alerts to homeowners' phones. These features can help prevent accidents and emergencies, making homes safer for individuals and families.

Reduced energy consumption: Smart home technology can also promote energy efficiency and reduce energy consumption. Automated systems and sensors can adjust the temperature, lighting, and appliances in a home based on occupancy and user preferences, which can result in lower energy bills and a reduced carbon footprint.

Economic growth: The growth of the smart home technology industry has also had a positive impact on the economy. The increased demand for smart home devices and services has created jobs and spurred innovation, leading to economic growth and development.

Environmental impact: As mentioned above, the energy-saving features of smart home technology can reduce energy consumption and help mitigate the impact of climate change. This has become an important consideration for individuals, governments, and businesses looking to reduce their carbon footprint and promote sustainability.

Summary

Smart home technology is designed to make daily living easier, more efficient, and more secure. It allows users to remotely control and monitor their home environment, saving time and energy and promoting safety and convenience.

Geriatric Smart Home Technology

"Join us on a journey to revolutionize the way we care for our aging loved ones. Through the pages of this book, you will discover the barriers and facilitators of geriatric smart home technology implementation and gain a deeper understanding of the potential impact it can have on their safety, health, and quality of life. Together, we can break down the barriers and pave the way for a brighter future for aging populations around the world. Let's embrace innovation and ensure that no one is left behind in the march towards progress!"

The convenience and efficiency of technology have made it a true blessing. Its primary function is to assist individuals with completing daily tasks that may require high amounts of physical stress. The elderly, in particular, view technology as a positive resource that enhances their lives, and technical issues do not significantly impact their daily routines. Smart home technology refers to the various communication technologies that allow appliances and objects in a home to be remotely managed, controlled, and automated. This concept can be applied to many household products, offering benefits for the elderly, disabled, and everyone in between.

Smart home technology can have a significant impact on different generations. Retirees can benefit from remote medical check-ups and access to basic services, which can be facilitated through smart home technology. Working professionals who own smart home products can save both time and money, allowing them to be more productive and contribute to a wealthier economy. The youth who grow up with smart home technology will have an advantage over those who do not, and this technology can indirectly contribute to longer life expectancies.

Summary

Technology has become a necessary and integral part of our lives, offering numerous benefits and solutions to various problems. Smart home technology is a prime example of how technology can be used to enhance the quality of life for individuals and society as a whole.

Overview of the Importance of Geriatric Smart Home Technology

"As the world's population ages, the use of technology to support aging in place has become increasingly important. Geriatric smart home technology has the potential to improve the quality of life for older adults by providing them with greater independence, safety, and convenience. However, the implementation of this technology is not without its challenges, and it is important to consider the barriers and facilitators to its adoption in order to ensure its success."

How can geriatric smart home technology improve the quality of life for older adults and support aging in place?

As the global population continues to age, the demand for innovative solutions to support older adults is on the rise. According to the World Health Organization, the number of people aged 60 years and older is expected to double by 2050, reaching 2 billion people worldwide. This demographic shift has led to a growing need for innovative solutions that can support older adults in maintaining their independence and quality of life. Geriatric smart home technology is one such solution, which has the potential to transform the way older adults live by enabling them to age in place, safely, independently, and comfortably. With the potential to revolutionize the way we care for older adults, geriatric smart home technology is more important now than ever before.

As we age, our physical and cognitive abilities can decline, making it challenging to complete even the simplest of daily tasks. This can have a significant impact on an individual's ability to live independently and can lead to a loss of autonomy and dignity. However, with the help of geriatric smart home technology, older adults can remain in their own homes and maintain their independence, without sacrificing their safety or quality of life.

The potential benefits of geriatric smart home technology are numerous and significant. First and foremost, this technology has the potential to enable older adults to live independently and safely in their own homes for longer periods of time, rather than having to move into assisted living facilities or nursing homes. This can be particularly important for older adults who have mobility or health issues, as well as those who live in rural areas or have limited access to transportation.

Geriatric smart home technology can also improve the quality of life for older adults by providing access to a range of support services and tools. For example, remote monitoring systems can allow healthcare providers to monitor vital signs and detect potential health issues before they become serious. Communication tools, such as video conferencing and messaging platforms, can help older adults stay connected to their caregivers and loved ones, reducing social isolation and loneliness. Home automation systems can simplify daily tasks, such as turning on lights or adjusting the temperature, making it easier for older adults to maintain their independence. They can also enhance older adults' quality of life by enabling them to stay connected with family and friends, pursue hobbies and interests, and access healthcare services more easily.

In addition to the potential benefits, the implementation of geriatric smart home technology also comes with various challenges that need to be addressed. One of the significant challenges is financial barriers, as the cost of implementing this technology can be high. Many older adults live on a fixed income, making it difficult for them to afford the upfront cost of these technologies. Furthermore, some insurance policies do not cover the cost of this technology, making it even more challenging for older adults to access it.

Technological challenges are also a significant barrier to the adoption of geriatric smart home technology. Many older adults may be less familiar with technology, making it difficult for them to learn and use these devices effectively. Additionally, these technologies may require a high level of technical knowledge to install and maintain, which can be a challenge for both older adults and their caregivers.

Societal and cultural barriers to adoption are also important to consider. Some older adults may be resistant to the idea of technology playing such a significant role in their daily lives. They may prefer traditional methods of care and support, such as in-person visits from caregivers or family members. Additionally, privacy concerns related to the collection and sharing of personal data may also make some older adults hesitant to adopt geriatric smart home technology.

Finally, the potential risks associated with geriatric smart home technology must be addressed. Privacy concerns related to the collection and sharing of personal data by these technologies may be a significant risk. Older adults may also experience social isolation if they rely too heavily on technology for communication and social interaction.

Summary

This book provides a comprehensive overview of the importance of geriatric smart home technology and its potential to improve the lives of older adults. It aims to highlight the challenges and opportunities associated with its implementation and provide insights into how stakeholders can work together to make this technology accessible and beneficial to older adults. The following chapters will delve deeper into specific issues related to geriatric smart home technology, such as the ethical and legal considerations, the role of caregivers, and the impact of technology on social isolation.

Current State of Geriatric Care: Challenges and Opportunities for Improvement

"Despite significant advancements in healthcare, the geriatric care system continues to face numerous challenges such as high costs, workforce shortages, and fragmented care. Technology, particularly geriatric smart home technology, has the potential to improve the quality of care, reduce healthcare costs, and enhance the independence and wellbeing of older adults."

What are the main challenges faced by the geriatric care system and how can technology help address them?

Older adults face a unique set of challenges that can make it difficult for them to live independently and lead a fulfilling life. The current state of geriatric care faces significant challenges in meeting the needs of the aging population. Despite advances in medical technology, older adults continue to experience gaps in care, resulting in suboptimal health outcomes and decreased quality of life. One of the main challenges is the lack of coordination between healthcare providers and social services, which can result in fragmented care and gaps in addressing the social determinants of health that impact older adults.

The growing demand for geriatric care presents significant challenges for healthcare providers and policymakers. There is a shortage of geriatric care providers, which can make it difficult for older adults to access the services they need. Furthermore, there is a need for more innovative solutions to meet the needs of older adults and enable them to live independently and safely.

Another challenge is the high cost of care, which can create financial barriers to accessing needed services and treatments. Many older adults are living on fixed incomes and may not have the financial resources to cover the cost of long-term care or expensive medical procedures. This can lead to delays in seeking care, which can exacerbate health conditions and result in poorer outcomes.

To address these challenges, there is a need for innovative solutions that integrate healthcare and social services to provide holistic care to older adults. Technology has the potential to play a significant role in improving the quality of care for older adults. Geriatric smart home technology has the potential to bring about numerous benefits for older adults and their caregivers. One of the primary benefits of geriatric smart home technology is that it can enable older adults to live independently for longer. By using monitoring devices, communication tools, and home automation systems, older adults can receive assistance with tasks such as medication management, meal preparation, and personal care. These technologies can also provide real-time information to caregivers about their loved ones' health and wellbeing, allowing for early intervention if necessary.

Another important benefit of geriatric smart home technology is that it can improve medication management. Many older adults struggle with managing multiple medications, which can lead to adverse health outcomes. By using smart medication dispensers and reminders, older adults can stay on top of their medication schedules and avoid missed doses or incorrect dosages.

Geriatric smart home technology can also help prevent falls, which are a leading cause of injury among older adults. By using sensors and monitoring systems, older adults can receive alerts if they are at risk of falling or have fallen. These systems can also provide real-time updates to caregivers and emergency responders, ensuring that older adults receive prompt assistance if needed.

Finally, geriatric smart home technology can enhance social connections for older adults. By using communication tools such as video conferencing and messaging apps, older adults can stay in touch with friends and family, reducing the risk of social isolation and loneliness.

One area where technology is making a significant impact is telehealth, which involves using technology to deliver healthcare services remotely. This can include video consultations with healthcare providers, remote monitoring of vital signs, and the use of mobile health applications to manage chronic conditions. Telehealth can help older adults access care more easily, reduce the need for hospitalizations, and improve medication adherence.

Mobile health applications are also becoming increasingly popular in geriatric care. These applications can provide older adults with tools to manage their health, such as medication reminders, exercise programs, and nutritional guidance. Mobile health applications can also be used to monitor health metrics, such as blood pressure and blood glucose levels, and provide feedback and guidance based on this information.

While technology has the potential to provide many benefits to older adults, it is important to consider the unique needs and limitations of this population. Geriatric smart home technology has the potential to improve the quality of life for older adults, there are also potential risks associated with these technologies. One significant concern is privacy, as these technologies involve the collection and sharing of sensitive personal information, such as medical data and daily activity patterns. If this information is mishandled or falls into the wrong hands, it could compromise the privacy and security of older adults.

Another potential risk associated with geriatric smart home technology is social isolation. While these technologies can help older adults live independently, they may also limit social interaction and lead to feelings of loneliness or disconnection. This is particularly concerning given the growing body of research indicating that social isolation can have negative impacts on physical and mental health in older adults.

It is essential to address these risks to ensure the successful implementation of geriatric smart home technology. Strategies to address these risks could include developing clear guidelines and regulations for the collection and use of personal data, providing education, adequate training and support to ensure that older adults and caregivers on how to use these technologies effectively & safely, and designing technologies with features that encourage social connection and interaction. It is important to design technology solutions with these limitations in mind and to provide adequate training and support to ensure that older adults can use the technology effectively.

Summary

It is critical to involve older adults and their caregivers in the development and implementation of these technologies to ensure that their needs and concerns are addressed. By addressing these risks and involving stakeholders in the development and implementation of geriatric smart home technology, we can work towards creating a safer and more connected future for older adults.

Understanding Geriatric Smart Home Technology: Key Features and Benefits

"Geriatric smart home technology has the potential to revolutionize the way we care for older adults by providing a safe, secure, and comfortable environment that promotes independence and enhances quality of life. From monitoring vital signs to automating household tasks, smart home technology offers a range of benefits that can help older adults age in place and maintain their dignity and autonomy."

What are the key features and benefits of geriatric smart home technology, and how can they improve the quality of life for older adults?

Geriatric smart home technology is a growing field that encompasses a wide range of devices and systems designed to support the health and wellbeing of aging populations. Geriatric smart home technology is a specialized type of home automation that uses sensors, cameras, and other connected devices to monitor and manage the home environment for the benefit of older adults. This technology is designed to support the health, safety, and independence of aging populations, allowing them to age in place while receiving the care and support they need.

At the heart of geriatric smart home technology are a variety of sensors that can be placed throughout the home to monitor activity, detect falls, and track vital signs like heart rate and blood pressure. These sensors can provide real-time information to caregivers or healthcare providers, allowing them to quickly respond to emergencies or changes in health status.

In addition to sensors, geriatric smart home technology may also include cameras and other monitoring devices that can provide visual feedback on the home environment. For example, a camera may be installed in the living room to allow caregivers to check in on a senior and ensure they are safe and comfortable. Other devices, like smart locks and motion sensors, can be used to detect when a senior leaves the house or returns home, providing peace of mind for caregivers and family members.

One of the key advantages of geriatric smart home technology is its ability to automate tasks and reduce the burden of caregiving. For example, a smart medication dispenser can be programmed to dispense pills at specific times, ensuring that a senior takes their medication as prescribed. Similarly, a smart thermostat can be programmed to adjust the temperature of the home based on a senior's preferences or schedule, reducing the need for manual adjustments.

In this chapter, we provide a comprehensive discussion of the different types of smart home technology available for geriatric care, highlighting their key features and benefits and explaining how they can improve the quality of life for seniors.

Monitoring Devices:

Monitoring devices are one of the most common types of smart home technology used in geriatric care. These devices can track vital signs, detect falls, and monitor medication adherence. They can also be programmed to send alerts to caregivers or healthcare providers in the event of an emergency. Some examples of monitoring devices include:

Wearable Sensors: Wearable sensors can track a variety of vital signs, including heart rate, blood pressure, and oxygen levels. They can also detect falls and other accidents, and send alerts to caregivers or emergency services if necessary.

Medication Dispensers: Medication dispensers can be programmed to dispense medications at specific times, ensuring that seniors take their medication as prescribed. They can also send alerts to caregivers or healthcare providers if a dose is missed.

Bed Sensors: Bed sensors can detect movement and changes in position, alerting caregivers or healthcare providers if a senior falls or experiences other health issues during the night.

Monitoring devices are an essential feature of geriatric smart home technology that can help improve the health outcomes of seniors and provide peace of mind for their caregivers. These devices can monitor vital signs such as blood pressure and heart rate, as well as detect falls and other emergencies. They can also monitor medication adherence, ensuring that seniors take their medications as prescribed.

One of the most significant advantages of monitoring devices is that they provide real-time information on the health status of seniors. This information can be transmitted to healthcare providers or family members, who can then take appropriate action if necessary. For example, if a senior's blood pressure or heart rate is outside of the normal range, healthcare providers can adjust their treatment plans accordingly. If a fall is detected, caregivers can be alerted immediately, allowing them to respond quickly and provide assistance.

Monitoring devices can also help seniors stay on top of their medication schedules. These devices can remind seniors to take their medications at the right time and in the correct dosage. They can also track medication adherence, providing caregivers with an accurate record of when medications were taken. This can be particularly beneficial for seniors who are managing multiple medications or who have memory impairments.

In addition to improving the health outcomes of seniors, monitoring devices can also provide peace of mind for their caregivers. Caregivers can monitor their loved one's health status remotely, reducing the need for in-person visits or check-ins. This can be particularly beneficial for long-distance caregivers or those who have other responsibilities, such as work or childcare.

It is important to note that monitoring devices are not a replacement for medical care. They are intended to supplement medical care and provide additional information to healthcare providers and caregivers. It is also important to ensure that seniors are comfortable with the use of monitoring devices and that their privacy and dignity are respected.

In conclusion, monitoring devices are a key feature of geriatric smart home technology that can help improve the health outcomes of seniors and provide peace of mind for their caregivers. These devices can monitor vital signs, detect falls and other emergencies, and ensure medication adherence. By providing real-time information on the health status of seniors, monitoring devices can help healthcare providers and caregivers take appropriate action if necessary.

Communication Tools:

Communication tools are another important type of smart home technology for geriatric care. These tools can help seniors stay connected with their loved ones and healthcare providers, reducing social isolation and improving mental health outcomes. Some examples of communication tools include:

Video Conferencing Systems: Video conferencing systems can be used to connect seniors with family members or healthcare providers who are not physically present. This can be particularly useful for seniors who have limited mobility or live far away from their loved ones.

Voice Assistants: Voice assistants like Amazon's Alexa or Google Home can provide seniors with access to information, entertainment, and communication tools without requiring them to use a computer or mobile device.

Messaging Apps: Messaging apps like WhatsApp or Skype can allow seniors to stay in touch with friends and family members, even if they are not physically present.

Communication tools are a vital feature of geriatric smart home technology. As seniors age, they may become isolated from their loved ones and healthcare providers, leading to social isolation and poor mental health outcomes. Communication tools such as video conferencing systems, voice assistants, and messaging apps can help seniors stay connected with their loved ones and healthcare providers, improving their quality of life and reducing the negative impacts of social isolation.

Video conferencing systems are particularly useful for seniors who are unable to leave their homes or who live far away from their loved ones. These systems allow seniors to see and speak with their loved ones in real-time, providing a sense of connection and reducing feelings of loneliness. Video conferencing can also be used for telehealth appointments, enabling seniors to receive medical care without leaving their homes.

Voice assistants such as Amazon's Alexa or Google Home can also be beneficial for seniors. These devices can be programmed to remind seniors to take their medications, schedule appointments, and provide information on their health conditions. Voice assistants can also provide entertainment and companionship, playing music or audiobooks and engaging in conversation with seniors.

Messaging apps such as WhatsApp or Facebook Messenger, like any other popular ones can also help seniors stay connected with their loved ones. These apps allow seniors to send and receive messages, photos, and videos, providing a way to communicate and share experiences with family and friends who may not be able to visit them in person.

In addition to improving social connectivity, communication tools can also help seniors manage their care and stay on top of their medication schedules. Voice assistants and messaging apps can provide reminders for medication schedules and appointments, ensuring that seniors do not miss important healthcare tasks. This can improve their health outcomes and reduce the risk of hospitalization or other health complications.

In conclusion, communication tools are an essential feature of geriatric smart home technology. Video conferencing systems, voice assistants, and messaging apps can help seniors stay connected with their loved ones and healthcare providers, reducing social isolation and improving mental health outcomes. These tools can also help seniors manage their care and stay on top of their medication schedules, improving their health outcomes and reducing the risk of hospitalization or other health complications.

Home Automation Systems:

Home automation systems are a third key type of smart home technology for geriatric care. These systems can control the lighting, temperature, and other environmental factors in the home, providing seniors with greater comfort and convenience. They can also help seniors to maintain their independence, allowing them to control their home environment even if they have mobility or cognitive impairments. Some examples of home automation systems include:

Smart Thermostats: Smart thermostats can be programmed to adjust the temperature of the home based on a senior's preferences or schedule. They can also be controlled remotely via a mobile app or voice assistant.

Smart Lighting: Smart lighting systems can be programmed to turn on and off automatically, or be controlled remotely via a mobile app or voice assistant. This can help seniors who have limited mobility or difficulty reaching light switches.

Smart Locks: Smart locks can be controlled remotely via a mobile app or voice assistant, allowing seniors to lock and unlock their doors without needing to physically interact with the lock.

Home automation systems are an essential aspect of geriatric smart home technology. These systems use sensors and other connected devices to control various environmental factors in the home, providing seniors with greater comfort and convenience. The systems can also help seniors to maintain their independence, even if they have mobility or cognitive impairments.

One of the key benefits of home automation systems is that they can control lighting. As seniors age, their vision may deteriorate, making it difficult for them to navigate their homes safely. Home automation systems can automatically adjust lighting levels, ensuring that seniors always have enough light to move around safely. They can also be programmed to turn on lights in response to movement, eliminating the need for seniors to fumble for light switches.

Temperature control is another essential feature of home automation systems. Seniors may have difficulty regulating their body temperature, making them susceptible to heat stroke and other heat-related illnesses. Home automation systems can be programmed to maintain a comfortable temperature throughout the day, ensuring that seniors remain safe and comfortable. They can also be used to control fans and air conditioning units, further improving comfort levels.

Home automation systems can also control other environmental factors in the home, such as door locks and window shades. This is particularly beneficial for seniors with mobility or cognitive impairments, who may struggle to move around the home or remember to lock doors and close blinds. By automating these processes, seniors can maintain their independence and feel more secure in their homes.

The benefits of home automation systems are not limited to seniors. Caregivers can also benefit from this technology, as it can help them to monitor their loved ones remotely. For example, if a senior forgets to turn off a stove, a caregiver can receive an alert through the home automation system and take appropriate action. This can provide peace of mind for caregivers, knowing that they can monitor their loved ones even when they are not physically present.

Overall, home automation systems are an essential aspect of geriatric smart home technology. They can improve the quality of life for seniors by providing greater comfort and convenience, as well as helping them to maintain their independence. They can also provide benefits for caregivers, allowing them to monitor their loved ones remotely and take appropriate action when necessary.

Energy Savings:

One of the important benefits of smart home technology is energy savings. Smart home technology offers a wide range of energy-saving features that can help older adults reduce their energy consumption and save money on their utility bills. One of the key energy-saving features of smart home technology is the ability to automate home systems such as heating, cooling, and lighting. This automation can be done through the use of sensors that detect when someone is in a room, or through a pre-set schedule that turns systems on and off at certain times of day.

For example, smart thermostats can adjust the temperature of a home based on occupancy and the time of day. These devices can learn a homeowner's patterns and preferences, adjusting the temperature automatically to optimize comfort while minimizing energy usage. Similarly, smart lighting systems can turn lights on and off automatically based on occupancy, or can be programmed to turn off lights in unoccupied rooms.

Another energy-saving feature of smart home technology is the ability to monitor and track energy usage. Smart meters and energy monitoring systems can provide detailed information on how much energy is being used in a home, and can identify areas where energy use can be reduced. This information can be used to make informed decisions about energy usage and to identify ways to reduce energy consumption.

In addition to saving energy and money, smart home technology can also promote safety and security. For example, smart smoke detectors can detect smoke and carbon monoxide and send alerts to homeowners' phones, enabling them to take action quickly. Smart security systems can also detect and alert homeowners to potential security threats, and can provide remote monitoring and control of home security systems.

Overall, the energy-saving features of geriatric smart home technology offer a wide range of benefits to older adults, including reduced energy consumption, cost savings, and increased safety and security. As you discuss the key features and benefits of geriatric smart home technology in your book, be sure to highlight the ways in which these technologies can promote energy efficiency and help older adults live more comfortable, safe, and sustainable lives.[5][6]

Summary

The different types of smart home technology available for geriatric care provide a range of benefits for seniors and their caregivers. From monitoring devices that can detect emergencies, to communication tools that can reduce social isolation, to home automation systems that can improve comfort and convenience, smart home technology has the potential to improve the quality of life for aging populations in a variety of ways.

Barriers to Geriatric Smart Home Technology Implementation: Financial, Technological, and Societal Challenges

"Geriatric smart home technology has the potential to improve the quality of life for older adults and reduce the burden on caregivers, but its implementation is not without challenges. Financial, technological, and societal barriers must be addressed to ensure that these technologies reach the people who need them the most."

What are the financial, technological, and societal challenges that are impeding the implementation of geriatric smart home technology, and what strategies can be used to overcome these barriers?

The implementation of geriatric smart home technology faces a number of barriers that can prevent its adoption and limit its impact on aging populations. In this chapter, we will discuss the key financial, technological, and societal challenges that must be overcome in order to successfully implement geriatric smart home technology. We will also provide real-world examples of these challenges and discuss strategies for overcoming them.

Financial Challenges:

One of the primary barriers to geriatric smart home

technology implementation is cost. Smart home technology can be expensive, and many seniors may not have the financial means to purchase and maintain these systems. Additionally, insurance companies and healthcare providers may not cover the costs of these technologies, making them unaffordable for many seniors.

The cost of geriatric smart home technology is a significant barrier to its adoption, particularly for seniors who may be on a fixed income or have limited financial resources. Smart home technology can be expensive, and the costs associated with purchasing and maintaining these systems can add up quickly. Additionally, insurance companies and healthcare providers may not cover the costs of these technologies, making them unaffordable for many seniors.

To overcome these financial challenges, it is important to identify funding opportunities and develop strategies for reducing the cost of geriatric smart home technology. One approach is to seek out grants or other funding sources that can be used to offset the costs of these technologies. This can include government grants or private foundation funding, as well as crowdfunding or other community-based fundraising initiatives.

Another approach is to partner with healthcare providers or insurance companies to cover the costs of geriatric smart home technology. For example, insurance companies may be willing to cover the cost of certain smart home devices if they can be shown to reduce healthcare costs or improve health outcomes. Similarly, healthcare providers may be willing to cover the cost of these technologies as part of a care plan for their patients.

Finally, developing more affordable technology solutions can also help overcome financial barriers to geriatric smart home technology implementation. This can include working with technology providers to develop low-cost alternatives to existing devices, or developing open-source technologies that can be customized and adapted to meet the needs of different populations.

In conclusion, financial challenges are a significant barrier to geriatric smart home technology implementation, but there are strategies that can be employed to overcome these challenges. By identifying funding opportunities, partnering with healthcare providers and insurance companies, and developing more affordable technology solutions, we can make smart home technology more accessible and affordable for aging populations.

Technological Challenges:

The technological challenges associated with geriatric smart home technology implementation are numerous and complex. One of the primary challenges is the issue of compatibility between different smart home devices. These devices may use different communication protocols or operate on different platforms, making it difficult to integrate them into a cohesive and effective system. This lack of interoperability can lead to frustration and confusion for seniors who may struggle to use multiple devices from different manufacturers.

Another key technological challenge is security. Geriatric smart home technology devices are often connected to the internet, which makes them vulnerable to hacking and cyber threats. This can result in the theft of sensitive data, or even compromise the safety and well-being of seniors who rely on these devices for assistance and support.

To overcome these technological challenges, it is important to develop standards for interoperability and security. Technology providers must work together to ensure that their devices are compatible with each other, and that they can communicate effectively to create a cohesive system. This can be achieved through the development of open standards and protocols, which can help ensure that devices from different manufacturers can communicate and work together seamlessly.

In addition, security measures must be implemented to protect sensitive data and prevent unauthorized access to smart home devices. This can include the use of strong encryption algorithms, authentication mechanisms, and firewalls to protect against cyber threats. Regular software updates and patches should also be implemented to address known vulnerabilities and improve overall security.

Overall, addressing the technological challenges associated with geriatric smart home technology implementation is essential for unlocking the full potential of these devices to improve the lives of seniors. By developing interoperability standards and implementing robust security measures, we can create a safe, reliable, and effective system that can help seniors age in place and maintain their independence.

Societal Challenges:

Societal challenges are a significant barrier to geriatric smart home technology implementation. The attitudes and perceptions about aging and technology can impact the willingness of seniors to adopt new technologies, limiting their potential benefits. Societal biases and stereotypes can also create barriers to the widespread adoption of geriatric smart home technology.

Many seniors may be resistant to adopting new technologies due to a lack of familiarity or concerns about privacy and autonomy. They may feel that technology is too complicated or not relevant to their needs, or worry that it may compromise their privacy or infringe on their independence. Additionally, there may be societal biases and stereotypes that limit the potential impact of geriatric smart home technology. For example, some may view aging as a time of decline and diminished capabilities, rather than as an opportunity for continued growth and development.

To overcome these societal challenges, it is important to promote public awareness and education about the benefits of geriatric smart home technology. This can include working with community organizations, healthcare providers, and advocacy groups to raise awareness about the potential benefits of these technologies. Providing clear and concise information about the benefits of geriatric smart home technology can help to dispel myths and misconceptions about aging and technology.

It is also important to involve seniors in the development and implementation of geriatric smart home technology, to ensure that their needs and concerns are being addressed. This can include consulting with seniors to understand their attitudes and perceptions about technology, as well as involving them in the design and testing of new technology solutions. By involving seniors in the development process, we can ensure that the technology meets their needs and preferences, and that it is designed to enhance their quality of life.

Finally, it is important to address societal biases and stereotypes that limit the potential impact of geriatric smart home technology. This can include challenging ageist attitudes and promoting positive images of aging and technology. By reframing the narrative around aging and technology, we can create a more inclusive and supportive environment for seniors to adopt and use geriatric smart home technology.

In conclusion, societal challenges are a significant barrier to geriatric smart home technology implementation. To overcome these challenges, it is important to promote public awareness and education, involve seniors in the development and implementation process, and challenge ageist attitudes and stereotypes. By doing so, we can unlock the potential of geriatric smart home technology to improve the lives of aging populations.

Real-World Examples:

There are many real-world examples of the barriers to geriatric smart home technology implementation. For example, a study published in the Journal of Gerontological Nursing found that the high cost of these technologies was a significant barrier to adoption. Similarly, a report by the National Institute on Aging identified concerns about privacy and security as a key challenge to implementing geriatric smart home technology.

The barriers to geriatric smart home technology implementation can be seen in many real-world examples. Here are a few examples:

High Cost: The cost of smart home technology can be a significant barrier for many seniors. In a study published in the Journal of Gerontological Nursing, researchers found that the high cost of these technologies was a significant barrier to adoption. The study surveyed older adults and found that many were unwilling or unable to pay for the cost of the technology. This is particularly concerning given that seniors often have fixed incomes and limited financial resources.

Privacy and Security Concerns: Privacy and security concerns are another barrier to geriatric smart home technology implementation. A report by the National Institute on Aging identified concerns about privacy and security as a key challenge to implementing geriatric smart home technology. Seniors may be hesitant to use these technologies if they believe that their personal information is at risk. This can include concerns about data breaches, hacking, and the use of personal data for advertising or other purposes.

Limited Technical Knowledge: Another real-world example of a barrier to geriatric smart home technology implementation is the limited technical knowledge of many seniors. Smart home technology can be complex, and seniors may struggle to understand how to use it. This can make it difficult for them to take advantage of the benefits of these technologies. In some cases, seniors may need additional training or support to help them understand how to use smart home technology effectively.

Lack of Compatibility: Smart home technology can also be limited by a lack of compatibility between devices. For example, if a senior has a smart thermostat from one manufacturer and a smart security system from another manufacturer, these devices may not be compatible with each other. This can make it difficult to create a cohesive and effective smart home system.

Limited Access to Technology: Finally, limited access to technology can be a significant barrier to geriatric smart home technology implementation. Seniors who live in rural or low-income areas may not have access to high-speed internet or the latest technology. This can limit their ability to take advantage of the benefits of smart home technology, even if they are willing and able to use it.

These real-world examples demonstrate the challenges that must be overcome in order to successfully implement geriatric smart home technology. By understanding these barriers and developing strategies to overcome them, we can unlock the potential of these technologies to improve the lives of aging populations. This includes addressing concerns about cost, privacy and security, technical knowledge, compatibility, and access to technology. By working together to overcome these challenges, we can create a future in which geriatric smart home technology is accessible and effective for all seniors.

Strategies for Overcoming Barriers:

To overcome these barriers, a number of strategies can be employed. For example, developing more affordable and accessible technology solutions can help address financial barriers, while implementing security measures and interoperability standards can help address technological challenges. Working with seniors to understand their concerns and needs can also help overcome societal barriers. To successfully implement geriatric smart home technology, it is important to employ strategies that can help overcome the barriers posed by financial, technological, and societal challenges. Here are some effective strategies that can be used to address these barriers:

Develop more affordable and accessible technology solutions: One of the most significant barriers to the adoption of geriatric smart home technology is the cost. To address this, it is important to develop more affordable and accessible technology solutions. This can be achieved by partnering with technology providers to develop cost-effective solutions, seeking out funding opportunities, and exploring alternative payment models, such as leasing or rental options.

Implement security measures and interoperability standards: Geriatric smart home technology can also face technological barriers, such as issues related to compatibility, interoperability, and security. To address these challenges, it is important to implement security measures and interoperability standards. This can be achieved by working with technology providers to ensure that their devices are compatible with each other, and implementing security measures to protect sensitive data.

Involve seniors in the development and implementation process: Seniors may be resistant to adopting new technologies due to a lack of familiarity or concerns about loss of privacy or autonomy. To address these societal barriers, it is important to involve seniors in the development and implementation process. This can be achieved by working with community organizations, healthcare providers, and advocacy groups to raise awareness about the potential benefits of these technologies and engaging seniors in the process to ensure that their needs and concerns are being addressed.

Provide education and training: Many seniors may not be familiar with the benefits and functions of geriatric smart home technology. To overcome this barrier, it is important to provide education and training on the use and benefits of these technologies. This can be achieved through workshops, seminars, and training sessions that are designed to familiarize seniors with the features and functionality of these technologies.

Partner with healthcare providers and insurance companies: Another effective strategy for overcoming financial barriers is to partner with healthcare providers and insurance companies. By working together, healthcare providers and insurance companies can help cover the costs of geriatric smart home technology for seniors, making it more accessible and affordable.

Overall, the barriers to geriatric smart home technology implementation are significant, but they can be overcome with the right strategies and approaches. By working to address financial, technological, and societal challenges, we can unlock the potential of smart home technology to improve the lives of aging populations.[7][8]

Summary

Overcoming the barriers to geriatric smart home technology implementation requires a multifaceted approach that addresses financial, technological, and societal challenges. By employing the right strategies and approaches, we can unlock the potential of smart home technology to improve the lives of aging populations. It is important to continue to work together as a community to develop innovative solutions that address the challenges and promote the benefits of geriatric smart home technology.

Facilitators of Geriatric Smart Home Technology Implementation: Strategies for Success

"To successfully implement geriatric smart home technology, it is essential to have a comprehensive strategy that takes into account funding opportunities, policy changes, and public awareness campaigns. By working together and utilizing these facilitators, we can create a more inclusive and accessible environment for older adults."

How can we ensure successful implementation of geriatric smart home technology in our communities?

This chapter focuses on the facilitators of geriatric smart home technology implementation and provides guidance on how to develop and implement successful strategies for geriatric smart home technology adoption. This chapter is important because it highlights the various factors that can help overcome barriers to implementation and ensure successful adoption of geriatric smart home technology.

Funding Opportunities:

One of the key facilitators of geriatric smart home technology implementation is funding opportunities. Financial resources are essential to support the development and implementation of smart home technology for geriatric care.

Governments, healthcare organizations, and philanthropic foundations can provide funding for research and development, pilot projects, and implementation initiatives. For example, grants can be provided to support the development of new smart home technology products, and subsidies can be offered to help seniors and their caregivers afford the cost of purchasing and installing these technologies.

Funding opportunities are a crucial facilitator of geriatric smart home technology implementation. These opportunities can provide financial support for research and development, as well as for the implementation of smart home technology in geriatric care settings.

There are several types of funding opportunities available for geriatric smart home technology implementation. One source of funding is government grants, which can be provided by federal, state, or local governments. These grants can be used to support research and development, pilot testing, and implementation of smart home technology in geriatric care settings.

Private investment is another source of funding that can be used to support geriatric smart home technology implementation. Private investors can provide funding for research and development, as well as for the implementation of smart home technology in geriatric care settings. Private investment can come from venture capitalists, angel investors, or other private investors.

Philanthropic organizations are also a potential source of funding for geriatric smart home technology implementation. Philanthropic organizations can provide grants or donations to support research and development, pilot testing, and implementation of smart home technology in geriatric care settings.

However, it is important to note that funding opportunities can be competitive, and organizations may need to submit a grant proposal or pitch to investors in order to secure funding. Organizations should be prepared to demonstrate the potential impact of their technology, as well as the feasibility and sustainability of their project.

In conclusion, funding opportunities are a crucial facilitator of geriatric smart home technology implementation. They provide financial support for research and development, pilot testing, and implementation of smart home technology in geriatric care settings. By leveraging these funding opportunities, organizations can drive innovation and development in the field of geriatric smart home technology and improve the quality of care for aging populations.

Policy Changes:

Another facilitator of geriatric smart home technology implementation is policy changes. Policymakers can play a crucial role in creating an environment that supports the adoption of geriatric smart home technology. For example, policies can be developed to incentivize the development and implementation of smart home technology for geriatric care, such as tax credits for companies that invest in research and development or regulatory frameworks that encourage the use of these technologies in healthcare settings. Policies can also be developed to ensure that the use of smart home technology for geriatric care is safe, secure, and ethical.

Policy changes are one of the facilitators of geriatric smart home technology implementation. Policies that support the development and adoption of smart home technology can help create a supportive environment for the implementation of these technologies in geriatric care settings.

There are several types of policy changes that can facilitate geriatric smart home technology implementation. One type of policy change is regulatory change. Regulatory changes can include changes to building codes or healthcare regulations that encourage the development and implementation of smart home technology. For example, building codes could require new construction or renovation of existing homes and care facilities to be designed to accommodate smart home technology. Healthcare regulations could be updated to include smart home technology as a reimbursable healthcare expense, which can incentivize healthcare providers to adopt the technology in their practices.

Another type of policy change is reimbursement policy changes. Reimbursement policy changes can include changes to Medicare or Medicaid reimbursement policies that provide financial incentives for healthcare providers to adopt smart home technology. For example, reimbursement policies could be updated to provide financial incentives for healthcare providers who use smart home technology to reduce readmission rates, improve patient outcomes, or improve overall care quality.

In addition to regulatory and reimbursement policy changes, public policy changes can also facilitate geriatric smart home technology implementation. Public policy changes can include public awareness campaigns that promote the benefits of smart home technology for geriatric care. These campaigns can help create demand for smart home technology, which can drive innovation and development. Public policy changes can also include changes to funding policies that provide financial support for the development and implementation of smart home technology.

In conclusion, policy changes are a facilitator of geriatric smart home technology implementation. Regulatory changes, reimbursement policy changes, and public policy changes can all help create a supportive environment for the development and adoption of smart home technology in geriatric care settings. By creating policies that support the development and adoption of smart home technology, policymakers can help improve the quality of care for aging populations and promote healthy aging in place.

Public Awareness:

Public awareness campaigns are also an important facilitator of geriatric smart home technology implementation. Raising awareness about the benefits of smart home technology for geriatric care can help overcome stigma and resistance to change. Public awareness campaigns can be developed to educate seniors, their caregivers, and healthcare providers about the potential benefits of smart home technology, such as improved safety, increased independence, and reduced healthcare costs. These campaigns can be delivered through various channels, such as social media, print and broadcast media, and community outreach programs.

Public awareness campaigns are an essential facilitator of geriatric smart home technology implementation. These campaigns can help educate the public and healthcare professionals about the benefits of smart home technology for geriatric care, and they can help create demand for these technologies, which can drive innovation and development.

One key goal of public awareness campaigns is to educate the public about the benefits of geriatric smart home technology. These benefits can include increased safety and security for elderly individuals, improved healthcare delivery and coordination, and increased independence and autonomy for aging populations. By highlighting these benefits, public awareness campaigns can help create demand for these technologies and encourage healthcare providers and policymakers to invest in their development and implementation.

Public awareness campaigns can also help address misconceptions and concerns about geriatric smart home technology. For example, some individuals may be concerned about privacy or data security issues, or may worry that the technology is too complicated or difficult to use. By addressing these concerns and providing clear and accurate information about the technology, public awareness campaigns can help build trust and confidence in geriatric smart home technology.

To be effective, public awareness campaigns must be targeted and tailored to specific audiences. This may involve creating different campaigns for different demographic groups, such as older adults, caregivers, or healthcare professionals. Campaigns should also be designed to reach individuals in a variety of settings, such as through social media, community events, or healthcare provider offices.

Another key aspect of public awareness campaigns is collaboration and partnerships. These campaigns should involve partnerships between healthcare providers, technology developers, and community organizations, as well as collaboration between different levels of government. By working together, these groups can leverage their strengths and resources to create more effective campaigns and reach a wider audience.

Overall, public awareness campaigns are an essential facilitator of geriatric smart home technology implementation. By educating the public about the benefits of these technologies, addressing concerns and misconceptions, and building demand and support, public awareness campaigns can help drive innovation and development in geriatric care and improve the quality of life for aging populations.

Strategies for Technology Adoption:

In addition to these facilitators, there are several strategies that can help organizations successfully adopt geriatric smart home technology.

To develop successful strategies for geriatric smart home technology adoption, it is important to consider the needs and preferences of seniors and their caregivers. Seniors and their caregivers should be involved in the development and implementation of smart home technology for geriatric care, to ensure that the technology meets their needs and is acceptable to them. Healthcare providers should also be involved in the development and implementation of smart home technology, to ensure that the technology is integrated into existing healthcare systems and workflows.

Geriatric smart home technology has the potential to significantly improve the quality of care for aging populations. However, successful adoption of these technologies requires careful planning and implementation. In this section, we will provide guidance on how to develop and implement successful strategies for geriatric smart home technology adoption.

Define Your Goals and Objectives: The first step in developing a successful strategy for geriatric smart home technology adoption is to define your goals and objectives. This includes identifying the specific problems you are trying to solve and the outcomes you hope to achieve through the use of smart home technology.

Engage Key Stakeholders: It is important to involve key stakeholders in the development and implementation of geriatric smart home technology. These stakeholders can include patients, caregivers, healthcare providers, and technology developers. Involving these stakeholders can help ensure that the technology meets the needs of its users and is designed with their input in mind.

Conduct a Needs Assessment: Before implementing geriatric smart home technology, it is important to conduct a needs assessment. This assessment should identify the specific needs and challenges of your target population, as well as any existing gaps in care that could be addressed through the use of smart home technology.

Identify and Evaluate Available Technology Solutions: There are a variety of geriatric smart home technology solutions available on the market. It is important to identify and evaluate these solutions based on their ability to meet your specific goals and objectives. This evaluation should include factors such as cost, ease of use, and compatibility with existing systems.

Pilot Test the Technology: Before implementing geriatric smart home technology on a larger scale, it is important to pilot test the technology. This can help identify potential issues or challenges before the technology is fully implemented, allowing organizations to make adjustments and improvements as needed.

Provide Training and Support: It is essential to provide training and support for patients and caregivers who will be using the technology. This can include training on how to use the technology effectively, as well as providing ongoing support and troubleshooting resources.

Monitor and Evaluate Outcomes: Finally, it is important to monitor and evaluate the effectiveness of the technology after it has been implemented. This can include monitoring patient outcomes, assessing the satisfaction of patients and caregivers, and evaluating the overall impact of the technology on healthcare delivery and costs.

Summary

This chapter provides guidance on the facilitators of geriatric smart home technology implementation, including funding opportunities, policy changes, and public awareness campaigns. It also emphasizes the importance of involving seniors, caregivers, and healthcare providers in the development and implementation of smart home technology for geriatric care, to ensure successful adoption and integration into existing healthcare systems. By considering these facilitators, stakeholders can work together to overcome barriers to implementation and improve the lives of seniors through the use of smart home technology.

Case Studies: Real-World Examples of Successful Implementation

"Real-world case studies provide valuable insights into the strategies and approaches that have led to successful geriatric smart home technology implementation. By examining these examples, we can identify best practices and learn from the challenges that were overcome to improve future implementation efforts."

What can we learn from successful case studies of geriatric smart home technology implementation and how can we apply these strategies to improve implementation in other contexts?

This section of the book as it provides real-world examples of successful geriatric smart home technology implementation, offering readers insight into the strategies and approaches that led to success. It can also help readers understand the challenges and obstacles that were overcome in these cases, and what can be learned from them. The case studies that are carefully selected to represent a diverse range of experiences and perspectives, highlighting examples of success across a variety of contexts. The case studies could be organized around the following themes:

Home Automation Systems:

Case Study 1: *Successful Implementation of Smart Home Automation in a Senior Living Community*

Introduction: The aim of this case study is to provide a real-world example of successful geriatric smart home technology implementation, highlighting the strategies and approaches that led to success. The focus will be on a senior living community that leveraged smart home technology to provide residents with greater autonomy and independence, while also improving staff efficiency and reducing healthcare costs.

Background: The senior living community in question is a 100-unit facility located in a suburban area. The community offers a range of care services, from independent living to assisted living and memory care. The community had been experiencing challenges in managing the care needs of its residents, particularly those with chronic conditions such as dementia, which require constant monitoring and intervention. Additionally, staff members were stretched thin, and the community was struggling to maintain high levels of care while also keeping costs down.

Implementation: The senior living community decided to implement a smart home automation system to help address these challenges. The system included a range of features, such as motion sensors, smart locks, and temperature sensors, which were integrated into the community's existing infrastructure. The system also included a mobile app that residents and staff members could use to control various aspects of the home environment.

To ensure successful implementation, the community took a phased approach, starting with a small pilot program in one of its assisted living units. During the pilot, the community worked closely with residents and staff members to identify any challenges and fine-tune the system. Based on the success of the pilot, the community gradually expanded the system to other units and care levels.

Results: The smart home automation system had a significant impact on the senior living community. Residents reported feeling more independent and in control of their living environment, which in turn led to higher levels of satisfaction and well-being. Staff members also reported feeling more efficient and effective in their care, as the system helped to streamline certain tasks and reduce the burden of constant monitoring.

Furthermore, the smart home automation system led to cost savings for the senior living community. By reducing the need for constant monitoring and intervention, the system helped to lower healthcare costs and reduce the risk of emergency hospitalizations.

Challenges and Obstacles: Despite its success, the implementation of the smart home automation system was not without its challenges. One of the main obstacles was the need to ensure that the system was accessible and user-friendly for all residents, including those with limited technological experience. The community addressed this by providing training and support to residents and staff members and ensuring that the system was intuitive and easy to use.

Another challenge was ensuring the security and privacy of resident data, particularly given the sensitive nature of the information being collected. The community addressed this by implementing robust data security measures and ensuring that resident data was only accessible to authorized personnel.

Lessons Learned: The successful implementation of the smart home automation system in this senior living community offers several important lessons for other care providers considering similar technology solutions. These include:

- Starting with a small pilot program to test the system and identify any challenges before scaling up implementation.

- Engaging residents and staff members throughout the implementation process to ensure buy-in and identify any issues or concerns.
- Ensuring that the system is accessible and user-friendly for all residents, regardless of technological experience.

Implementing robust data security measures to protect resident data.

Conclusion: The implementation of a smart home automation system in this senior living community offers a real-world example of successful geriatric smart home technology implementation. By leveraging smart home technology, the community was able to provide residents with greater autonomy and independence, improve staff efficiency, and reduce healthcare costs. The successful implementation was achieved by taking a phased approach, engaging residents and staff members, and addressing challenges such as accessibility and data security.

Case Study 2: *Implementing Smart Home Technology to Improve Senior Living*

Introduction: The use of smart home technology in geriatric care settings has become increasingly popular in recent years. One senior living community in California successfully leveraged smart home technology to provide its residents with greater autonomy and independence, while also improving staff efficiency and reducing healthcare costs. This case study will examine the strategies and approaches that led to success, as well as the challenges and obstacles that were overcome in the implementation process.

Method: The senior living community selected for this case study implemented a home automation system that included smart thermostats, smart lighting, and smart security cameras. The system was designed to give residents greater control over their living environment, while also improving staff efficiency and reducing healthcare costs.

To evaluate the success of the system, data was collected on resident satisfaction levels, staff efficiency, and healthcare costs before and after implementation. Interviews were also conducted with staff members and residents to gain a better understanding of the impact of the technology on their daily lives.

Results: The implementation of the smart home technology had a significant positive impact on both residents and staff. Residents reported increased satisfaction with their living environment and greater feelings of independence and control. Staff reported greater efficiency in their daily tasks, as well as reduced healthcare costs.

The success of the implementation was due in large part to the careful planning and communication strategies used by the senior living community. Prior to implementation, staff members were trained on the new system and were provided with resources to support residents in using the technology. Ongoing support was also provided to residents to ensure that they were able to fully utilize the system.

Challenges: One of the primary challenges in implementing the technology was ensuring that all residents were able to fully utilize the system. While many residents were comfortable with the technology, others required additional support and training. This challenge was addressed through ongoing support and training, as well as the availability of staff members to assist residents as needed.

Conclusion: Overall, the implementation of the smart home technology was highly successful in improving the lives of residents and staff in the senior living community. By providing residents with greater autonomy and independence, while also improving staff efficiency and reducing healthcare costs, the technology has become an integral part of the community's approach to geriatric care.

Wearable Technology:

Case Study: *Successful Implementation of Wearable Technology in Geriatric Care Settings*

Introduction: Wearable technology has the potential to transform the way we deliver care to elderly patients. With the rise of smart devices, it is now possible to monitor vital signs and alert caregivers in real-time. In this case study, we will examine the implementation of wearable technology in a geriatric care setting and highlight the strategies and approaches that led to its success. We will also discuss the challenges and obstacles that were overcome in this case and what can be learned from them.

Background: A senior living community in California implemented wearable technology devices to monitor vital signs and alert caregivers in the event of an emergency. The devices were designed to track heart rate, blood pressure, oxygen saturation, and respiratory rate. If any of these metrics fell outside of the normal range, an alert would be sent to the care team, who could then respond quickly.

Strategies and Approaches: The implementation of wearable technology in this geriatric care setting was successful due to several key strategies and approaches. First, the care team involved the patients and their families in the implementation process. This helped to ensure that the devices were comfortable and easy to use. Second, the care team provided comprehensive training to the patients and their families to ensure that they were comfortable using the devices. Third, the care team monitored the data collected by the devices closely and used it to adjust care plans and interventions.

Challenges and Obstacles: Despite the success of the implementation, there were some challenges and obstacles that needed to be overcome. One of the biggest challenges was resistance from some patients and families who were skeptical of the devices. The care team addressed this by providing education and support to these individuals. Another challenge was ensuring that the devices were worn consistently and correctly. The care team addressed this by providing ongoing monitoring and support to patients and families.

Lessons Learned: The implementation of wearable technology in this geriatric care setting provided several important lessons. First, involving patients and families in the implementation process is critical to success. Second, comprehensive training and support are essential to ensure that patients and families are comfortable using the devices. Third, ongoing monitoring and support are necessary to ensure that the devices are being used consistently and correctly.

Conclusion: The successful implementation of wearable technology in this geriatric care setting has the potential to improve patient outcomes and reduce healthcare costs. By tracking vital signs and alerting caregivers in real-time, patients can receive timely interventions and avoid hospitalizations. The strategies and approaches used in this case can serve as a model for other geriatric care settings looking to implement wearable technology.

Remote Monitoring:

Case Study: *Remote Monitoring for Geriatric Care*

Introduction: Remote monitoring technology has been increasingly used in geriatric care settings to help seniors manage chronic conditions more effectively. This case study highlights a real-world example of successful implementation of remote monitoring technology in a senior living community, focusing on the strategies and approaches that led to success, as well as the challenges and obstacles that were overcome.

Background: The senior living community in question had a high population of residents with chronic conditions, and medication adherence was a major concern for healthcare providers. The community had previously relied on in-person medication checks, which were time-consuming and costly. In addition, healthcare providers were concerned that some residents were not adhering to their medication regimens, which increased the risk of hospitalizations and emergency room visits.

Implementation: The senior living community implemented a remote monitoring system to track medication adherence among residents with chronic conditions. The system included wireless medication dispensers that were installed in residents' homes, as well as a web-based platform that allowed healthcare providers to monitor medication adherence in real-time. The system also included alerts that were sent to healthcare providers if a resident missed a medication dose.

Results: The implementation of the remote monitoring system resulted in a significant improvement in medication adherence among residents with chronic conditions. Healthcare providers were able to monitor medication adherence in real-time, which allowed them to intervene quickly if a resident missed a dose. The system also reduced the need for in-person medication checks, which saved time and reduced healthcare costs. In addition, the system helped seniors stay healthy and reduced the risk of hospitalizations and emergency room visits.

Challenges and Obstacles: The implementation of the remote monitoring system was not without challenges. Initially, some residents were hesitant to use the new technology, and there was a learning curve for both residents and healthcare providers. In addition, the cost of the technology was a concern for the senior living community. However, these challenges were overcome through education and training, and the benefits of the technology outweighed the initial costs.

Conclusion: The successful implementation of remote monitoring technology in this senior living community demonstrates the potential benefits of using technology to manage chronic conditions in geriatric care settings. By monitoring medication adherence in real-time, healthcare providers were able to intervene quickly if necessary, which helped seniors stay healthy and reduced the risk of hospitalizations and emergency room visits. While there were challenges and obstacles along the way, the benefits of the technology ultimately led to its successful implementation.

Communication Tools:

Case Study: *Video Conferencing Technology in Geriatric Care Settings*

Introduction: Loneliness and social isolation are significant problems among the elderly population, with potentially negative consequences for their health and well-being. One solution to this problem is the use of video conferencing technology to help seniors stay connected with family members and healthcare providers, even if they are unable to leave their homes. This case study examines the successful implementation of video conferencing technology in a geriatric care setting and highlights the strategies and approaches that led to success.

Background: The geriatric care setting in question is a senior living community located in a suburban area. The community provides housing and healthcare services to seniors who require some level of assistance with daily living activities. Prior to the implementation of video conferencing technology, many residents struggled with social isolation and loneliness, as they were unable to leave their homes due to mobility issues or other health concerns.

Implementation Strategy: To address the issue of social isolation and loneliness, the senior living community implemented a video conferencing system that allowed residents to connect with family members and healthcare providers. The system was designed to be user-friendly, with large buttons and simple navigation menus. The staff also provided training and support to residents to ensure that they were comfortable using the technology.

Challenges and Obstacles: One of the main challenges faced during the implementation of the video conferencing system was ensuring that residents had access to high-speed internet. In some cases, the community had to upgrade its internet infrastructure to ensure that the system would work properly. Another challenge was ensuring that family members were available and willing to participate in video calls with their loved ones.

Results: The implementation of video conferencing technology had a significant positive impact on the senior living community. Residents reported feeling more socially connected and less isolated, and many reported that they were able to communicate more frequently with family members and healthcare providers. The technology also allowed residents to attend virtual events and activities, further increasing their social engagement. Overall, the implementation of video conferencing technology was considered a success by both residents and staff.

Lessons Learned: The success of this implementation was due in part to the community's focus on user-friendly technology and staff support and training. Additionally, the community worked closely with family members to ensure their participation in video calls. One important lesson learned was the need to ensure that residents have access to high-speed internet to ensure that the system works properly.

Conclusion: The successful implementation of video conferencing technology in this geriatric care setting demonstrates the potential of technology to address issues of social isolation and loneliness among seniors. By focusing on user-friendly technology and staff support and training, this community was able to achieve significant positive outcomes for its residents.

Summary

This chapter provide readers with valuable insights into the real-world implementation of geriatric smart home technology. This information can help readers understand the factors that contribute to successful implementation of geriatric smart home technology, as well as the potential roadblocks that they may encounter. By examining successful case studies, readers can gain a deeper understanding of the potential benefits of this technology, as well as the strategies and approaches that are most likely to lead to success.

Ethical Considerations: Balancing Autonomy and Privacy with Technological Advancements

"As the use of geriatric smart home technology continues to grow, it is essential that we consider the ethical and legal implications of these advancements. Balancing the benefits of improved care and independence with the need to protect privacy and autonomy requires careful consideration and a commitment to best practices."

How can we ensure that the implementation of geriatric smart home technology respects the autonomy and privacy of older adults, while also promoting their health and well-being?

The implementation of geriatric smart home technology has the potential to improve the quality of life for seniors by providing them with greater independence, safety, and comfort. However, it also raises several ethical considerations that need to be addressed to ensure that the technology is used in an appropriate and responsible manner. It is therefore essential to establish clear guidelines for the use of geriatric smart home technology that prioritizes seniors' autonomy and privacy while also maximizing the benefits of this technology. In this discussion, we will explore the ethical considerations related to geriatric smart home technology implementation and provide guidance on how to navigate these ethical considerations.

One of the primary ethical considerations related to geriatric smart home technology implementation is the issue of autonomy. Autonomy is the principle of respecting an individual's right to make decisions for themselves, even if those decisions are not in their best interest. When it comes to geriatric care, autonomy is particularly important, as many seniors may have limited control over their lives due to physical or cognitive impairments.

Geriatric smart home technology has the potential to enhance seniors' autonomy by allowing them to remain in their homes and maintain their independence for longer. Seniors have the right to make decisions about their care, and the implementation of geriatric smart home technology should not infringe upon this right. It is important to involve seniors in the decision-making process and obtain their informed consent before implementing the technology. This involves providing clear information about the benefits and risks of the technology, as well as respecting seniors' preferences and values. However, this technology also raises ethical concerns related to privacy and surveillance.

Another ethical considerations related to geriatric smart home technology implementation is privacy. Geriatric smart home technology typically involves the use of sensors, cameras, and other monitoring devices to track seniors' movements and activities. This data can be sensitive and potentially intrusive, which raises concerns about privacy and surveillance. It is important to establish clear guidelines for data collection, storage, and sharing to ensure that seniors' privacy is protected. This may involve implementing data encryption, access controls, and other security measures.

Another ethical consideration is the potential for discrimination. Geriatric smart home technology may exacerbate existing inequalities in healthcare by favoring certain groups over others. For example, seniors who have access to the technology may receive better care than those who do not. To address this concern, it is important to ensure that the technology is accessible to all seniors, regardless of their socioeconomic status or other personal characteristics.

Finally, there is the issue of accountability. The implementation of geriatric smart home technology raises questions about who is responsible for the quality of care provided. This requires a clear understanding of the roles and responsibilities of different stakeholders, including healthcare providers, caregivers, and technology vendors. It is important to establish accountability mechanisms to ensure that all parties involved are held accountable for their actions.

To navigate these ethical considerations, it is important to establish clear guidelines for the use of geriatric smart home technology. This involves developing policies and procedures that prioritize seniors' privacy, autonomy, and accessibility, as well as establishing accountability mechanisms to ensure that these policies are enforced. It is also important to involve seniors in the decision-making process and obtain their informed consent before implementing the technology.

To address these ethical considerations, it is important to establish clear guidelines for the use of geriatric smart home technology. These guidelines should prioritize seniors' autonomy while also ensuring that their privacy and security are protected. Some of the key ethical considerations that should be addressed include:

Informed Consent: Seniors should have the right to make informed decisions about the use of geriatric smart home technology. This requires providing clear and concise information about the benefits and risks of the technology, as well as obtaining their explicit consent before implementing it.

Privacy and Security: Seniors' privacy and security should be protected at all times. This requires implementing appropriate security measures, such as encryption and password protection, and limiting access to sensitive data to authorized personnel only.

Data Ownership: Seniors should retain ownership of their data and have the right to access and control it. This requires implementing clear policies regarding data ownership and providing seniors with access to their data in a user-friendly format.

Transparency: Geriatric smart home technology should be transparent and open to scrutiny ensuring that it is used in a responsible and ethical manner. Clear information about how the technology works, what data is collected, and how that data is used should be provided to seniors to ensure they understand the implications of using this technology.

Fairness: Geriatric smart home technology should be implemented in a fair and equitable manner. This requires ensuring that all seniors have access to the technology, regardless of their socio-economic status or other personal characteristics.

Summary

It is crucial to establish clear guidelines for the use of geriatric smart home technology that prioritize seniors' autonomy and privacy while also maximizing the benefits of this technology. Addressing ethical considerations related to geriatric smart home technology will ensure that this technology is used in a responsible and ethical manner, benefiting seniors and society as a whole. By balancing autonomy and privacy with technological advancements, we can create a more equitable and just world for seniors.

Future of Geriatric Smart Home Technology: Innovations and Trends

"The possibilities for geriatric smart home technology are endless, and we are just scratching the surface of what is possible. As advancements continue to emerge, we can expect to see even more personalized and responsive care for the elderly, helping them to live more independently and with greater ease."

What kind of new geriatric smart home technologies will emerge in the future, and how will they improve the quality of life for aging individuals?

As the population ages, the demand for geriatric care is increasing rapidly. The future of geriatric smart home technology looks promising, with new innovations and trends emerging that have the potential to transform the way we care for aging populations. This chapter discusses the emerging technologies and trends that are likely to shape the future of geriatric care. In this chapter, we will explore some of the most exciting and promising innovations in geriatric smart home technology and discuss their potential impact on the aging population.

Artificial Intelligence (AI) and Machine Learning:

Artificial intelligence (AI) and machine learning have the potential to revolutionize the field of geriatric care by enabling more efficient and effective care for the aging population. In this section, we will explore the future of geriatric smart home technology with a focus on AI and machine learning.

AI and machine learning have the potential to greatly enhance the capabilities of smart home systems. With AI and machine learning, smart home systems will be able to learn the habits and preferences of seniors and adjust their settings accordingly. For example, the system could learn when a senior typically wakes up and adjust the lighting and temperature accordingly. These adjustments can greatly improve the quality of life for seniors by providing a more comfortable living environment.

Another potential use of AI and machine learning is in the early detection of health issues. Smart home systems can monitor a senior's behavior and activity levels, and use AI and machine learning algorithms to detect any changes that could indicate the onset of a health issue. This information can then be relayed to caregivers or healthcare providers, allowing for more timely and effective treatment.

AI and machine learning can also be used to analyze data from wearable devices, such as fitness trackers and smart watches. This data can provide valuable insights into a senior's health and wellness, allowing for more personalized care. For example, if a senior's wearable device indicates a decrease in physical activity, the smart home system could encourage the senior to engage in more physical activity or alert caregivers to any potential health issues.

In addition, AI and machine learning can be used to create predictive models that can identify potential health issues before they occur. For example, a smart home system could use AI and machine learning algorithms to analyze a senior's health data and predict the likelihood of falls or other health issues. This information can then be used to develop targeted interventions that can prevent these issues from occurring.

Overall, the use of AI and machine learning in geriatric smart home technology has the potential to greatly improve the quality of life for seniors and enhance the capabilities of caregivers and healthcare providers. By providing more personalized and proactive care, these innovations can enable seniors to live independently and safely in their own homes for longer. However, it is important to ensure that these technologies are developed with a focus on ethical and privacy considerations to ensure that they are safe and beneficial for all users.

Wearable Technology in Conjunction with Smart Home Systems:

The use of wearable technology in conjunction with smart home systems is a significant innovation and trend in the future of geriatric smart home technology. Wearable devices, such as fitness trackers and smartwatches, are already widely used by seniors to monitor their physical activity and health metrics. The integration of wearable technology with smart home systems has the potential to provide a more comprehensive and personalized approach to geriatric care.

One of the primary benefits of using wearable technology in conjunction with smart home systems is the ability to gather and analyze data on a senior's health and wellness. Wearable devices can track a range of health metrics, including heart rate, blood pressure, sleep patterns, and physical activity. This data can then be integrated into smart home systems to provide a more holistic view of a senior's health and wellness.

Smart home systems can use this data to adjust settings and alerts based on a senior's health status. For example, if a senior's heart rate is abnormally high or their sleep patterns are disrupted, the smart home system could adjust the lighting and temperature to create a more relaxing and calming environment. The system could also alert caregivers or emergency services if necessary.

Another potential benefit of using wearable technology in conjunction with smart home systems is the ability to provide personalized and tailored care. Wearable devices can provide data on a senior's specific health needs and preferences, which can then be used to customize their smart home system. For example, if a senior has mobility issues, the system could adjust the lighting and temperature based on their movement patterns to reduce the risk of falls.

The use of wearable technology in conjunction with smart home systems also has the potential to improve the overall quality of life for seniors. By providing personalized and tailored care, seniors can feel more comfortable and secure in their own homes. The system can also provide reminders for medication and appointments, reducing the risk of missed or forgotten appointments.

In conclusion, the use of wearable technology in conjunction with smart home systems is an exciting innovation and trend in the future of geriatric smart home technology. The integration of wearable devices with smart home systems has the potential to provide a more comprehensive and personalized approach to geriatric care. By gathering and analyzing data on a senior's health and wellness, smart home systems can adjust settings and alerts to provide a more comfortable and secure environment. This innovation has the potential to improve the overall quality of life for seniors and transform the field of geriatric care.

Virtual and Augmented Reality:

Virtual and augmented reality (VR/AR) are rapidly becoming important components of many modern technologies. VR creates a completely immersive digital environment that can be interacted with using specialized equipment, while AR overlays digital content onto the real world. Both of these technologies have the potential to revolutionize geriatric care and improve the quality of life for seniors living in smart homes.

One major advantage of VR and AR is that they can help seniors stay mentally and physically active. For example, virtual exercise programs can be designed specifically for seniors, allowing them to exercise safely and comfortably in their own homes. VR and AR can also be used to create games and activities that seniors can participate in, which can help reduce feelings of isolation and depression.

Another potential use for VR and AR is in memory care. These technologies can create digital environments that are familiar and comforting for seniors with memory loss, helping to reduce anxiety and confusion. VR and AR can also be used to simulate past experiences, which can help seniors with memory loss to maintain a sense of connection to their past.

In addition to improving mental and physical health, VR and AR can also improve safety in smart homes. For example, VR can be used to simulate potential hazards in a senior's home, such as loose rugs or cluttered walkways. This can help seniors and their caregivers identify and mitigate potential hazards before they lead to falls or other accidents.

Another potential use for VR and AR is in telemedicine. With these technologies, doctors and caregivers can remotely monitor a senior's health and provide real-time feedback and support. This can be especially beneficial for seniors who live in rural or remote areas, where access to healthcare services may be limited.

Despite the many potential benefits of VR and AR in geriatric smart home technology, there are also some challenges to consider. For example, many seniors may be uncomfortable or unfamiliar with these technologies, and may require additional training and support to use them effectively.

Additionally, some seniors may have physical limitations that prevent them from using certain VR and AR equipment.

Overall, the use of VR and AR in geriatric smart home technology is an exciting and rapidly evolving field. These technologies have the potential to improve mental and physical health, enhance safety, and improve access to healthcare services for seniors. As the technology continues to advance, it is likely that we will see even more innovative and impactful uses for VR and AR in the field of geriatric care.

Voice-activated Technology:

Voice-activated technology is one of the most exciting and promising trends in the future of geriatric smart home technology. Voice-activated assistants like Amazon's Alexa and Google Home are already becoming more and more prevalent in homes around the world, and the potential applications for seniors are particularly compelling.

One of the main benefits of voice-activated technology is that it enables seniors to control their home environments without having to physically interact with devices. This is particularly beneficial for those with mobility issues, such as arthritis or Parkinson's disease. By simply speaking a command, seniors can adjust the lighting, temperature, and other settings in their homes.

Voice-activated technology can also be integrated with other smart home systems, such as security cameras and motion sensors. This enables seniors to monitor their homes more effectively and respond to potential security threats in real-time.

In addition to these practical benefits, voice-activated technology can also have significant social and emotional benefits for seniors. Loneliness and isolation are major issues for many seniors, particularly those who live alone.

Voice-activated assistants can provide companionship and entertainment, playing music, reading books, and even telling jokes and stories.

As voice-activated technology continues to evolve, it is likely to become even more sophisticated and powerful. For example, it may be possible to use voice commands to order groceries, schedule appointments, or even control medical devices. This could be particularly beneficial for seniors with chronic health conditions, who may have difficulty leaving their homes to access medical care.

However, there are also some potential concerns associated with voice-activated technology. One major issue is privacy and security. Voice-activated assistants are always listening, which means that they may record sensitive information without the user's knowledge or consent. There are also concerns about the potential for hacking and unauthorized access to personal data.

Despite these concerns, the potential benefits of voice-activated technology for seniors are significant. By enabling seniors to live more independently, safely, and comfortably in their own homes, voice-activated technology has the potential to improve the quality of life for millions of seniors around the world.

In conclusion, voice-activated technology is one of the most exciting and promising trends in the future of geriatric smart home technology. By enabling seniors to control their home environments more easily, providing companionship and entertainment, and potentially even enabling greater access to medical care, voice-activated technology has the potential to transform the field of geriatric care. While there are some potential concerns that need to be addressed, the overall impact of this technology on the aging population is likely to be overwhelmingly positive.

Interconnectivity:

Interconnectivity is a key trend in the future of geriatric smart home technology. Interconnectivity refers to the ability of different devices and systems to communicate and exchange data with each other, creating a seamless and integrated experience for users.

One example of interconnectivity in geriatric smart home technology is the integration of smart home systems with electronic health records (EHRs) and other healthcare systems. This will allow healthcare providers to access and monitor a senior's health data remotely, enabling more proactive and effective care. For example, if a senior's EHR data indicates that they have not been taking their medication as prescribed, their smart home system could alert them and remind them to take their medication.

Another example of interconnectivity is the integration of smart home systems with telemedicine platforms. Telemedicine allows seniors to access healthcare services remotely, without having to leave their homes. Smart home systems could be used to facilitate telemedicine appointments by providing the necessary equipment and technology, such as video conferencing capabilities.

Interconnectivity can also be achieved through the use of application programming interfaces (APIs), which allow different systems and devices to communicate with each other. APIs can be used to integrate smart home systems with other devices and platforms, such as wearable technology and home automation systems.

The potential impact of interconnectivity on geriatric care and the aging population is significant. By enabling healthcare providers to access and monitor a senior's health data remotely, interconnectivity can improve the quality and efficiency of care. It can also enable seniors to access healthcare services more easily and conveniently, without having to leave their homes.

Interconnectivity can also improve the overall quality of life for seniors by creating a more seamless and integrated experience. For example, if a senior has a fall and their smart home system detects it, the system could automatically alert their healthcare provider and family members, reducing the response time and potentially preventing further injury.

However, there are also potential challenges and concerns associated with interconnectivity. One concern is privacy and security, as interconnectivity can increase the risk of data breaches and unauthorized access to sensitive information. It will be important for smart home systems to have robust security measures in place to protect the privacy and security of users.

In conclusion, interconnectivity is a key trend in the future of geriatric smart home technology. By enabling different devices and systems to communicate and exchange data with each other, interconnectivity can improve the quality and efficiency of geriatric care and enhance the overall quality of life for seniors. However, it is important to address concerns around privacy and security to ensure that the benefits of interconnectivity are fully realized.

Robotics

Robotics is another area of geriatric smart home technology that is showing great promise. Robots can assist with tasks such as cleaning, cooking, and even companionship. They can also be programmed to perform specific tasks related to health monitoring and care, such as taking blood pressure readings or reminding patients to take their medications.

In recent years, the field of geriatric care has been revolutionized by the emergence of smart home technology.

Smart home technology has enabled seniors to live independently and safely in their own homes for longer, while also providing caregivers with valuable insights into their health and wellness. Looking to the future, there are a number of innovations and trends that are likely to emerge in the field of geriatric smart home technology.

One major trend that is likely to emerge in the coming years is the development of more personalized and tailored solutions. Rather than relying on a one-size-fits-all approach, smart home systems will be able to be customized to meet the specific needs and preferences of each senior. For example, the system could learn a senior's preferred lighting and temperature settings, or the type of music they like to listen to, and adjust accordingly.

This level of personalization will not only improve the quality of life for seniors, but will also enable them to live independently for longer. By tailoring the smart home system to meet their individual needs, seniors will be able to perform tasks and activities that might otherwise be difficult or impossible.

The potential impact of these innovations on geriatric care and the aging population is significant. By enabling seniors to live independently for longer, personalized smart home solutions can improve their quality of life and reduce the burden on caregivers. Additionally, by providing caregivers with real-time data on a senior's health and wellness, smart home systems can enable more proactive and effective care, reducing the likelihood of hospitalizations or emergency room visits.

Summary

The future of geriatric smart home technology is full of exciting innovations and trends. The integration of AI, wearable devices, and other emerging technologies is likely to lead to more personalized and targeted support for aging populations. These innovations have the potential to improve the quality of life for older adults, reduce healthcare costs, and allow individuals to age in place with greater independence and dignity. By staying up-to-date with the latest developments in smart home technology, caregivers can provide seniors with the support and care they need to age gracefully and comfortably in their own homes. However, it is important to ensure that these technologies are developed and implemented in a way that prioritizes patient safety, privacy, and autonomy. By doing so, we can unlock the full potential of geriatric smart home technology and provide better care for our aging population.

Geriatric Smart Home Technology in India: Current Status, Future Prospects, and Challenges

In India, there is a growing need for geriatric smart home technology due to the rapidly increasing aging population. According to the United Nations, India is expected to have over 340 million people aged 60 and above by 2050, making it one of the largest aging populations in the world. In recent years, India has witnessed significant developments in the field of geriatric smart home technology. With a growing aging population, there is an increasing need for innovative solutions that can enhance the quality of life for older adults and support aging in place.

Present Status and Infrastructure:

India's current status in terms of geriatric smart home technology implementation is still in its nascent stage. However, several initiatives and projects have been undertaken in recent years to promote the adoption of this technology. The Indian government's Smart Cities Mission aims to create sustainable and livable cities with a focus on technology-driven solutions. Additionally, several private organizations have begun developing and implementing smart home technology for older adults.

Framework and Regulations:

Currently, there are no specific regulations or frameworks in India for geriatric smart home technology implementation. This lack of regulatory clarity may hinder the development of this technology and discourage investment in the sector. However, several government agencies and private organizations are working to establish guidelines and standards for the development and implementation of this technology.

Awareness and Advocacy:

Awareness and advocacy for the benefits of geriatric smart home technology among the Indian population is still low, and many older adults and their families may not even be aware of its existence. However, several advocacy organizations and initiatives are working to promote awareness of the potential benefits of this technology for older adults.

Future Prospects and Challenges:

The future prospects for geriatric smart home technology in India are promising. With a growing aging population and an increasing need for innovative solutions to support independent living, there is a significant potential for this technology in India. However, challenges such as affordability, lack of infrastructure, and cultural barriers to technology adoption must be addressed to ensure the successful implementation of this technology.

Education and Research:

Limited education and training opportunities for healthcare professionals and caregivers in the use of geriatric smart home technology. This may result in a reluctance to adopt and utilize the technology to its full potential. India is home to several institutions and universities conducting research on geriatric smart home technology. However, there is a need for increased investment in research and education to further advance the development and implementation

However, there are some promising developments in the field. The Indian government has taken steps to encourage the development and adoption of assistive technologies, including geriatric smart home technology. In 2021, the Ministry of Social Justice and Empowerment launched a scheme to provide assistive devices and technologies to senior citizens, including smart home solutions.

There are also some private sector initiatives focused on the development of geriatric smart home technology, such as the startup Niramai which has developed a smart mirror that can detect early-stage breast cancer in women.

Moving forward, it will be important for India to invest in research and education to promote awareness and understanding of the benefits of geriatric smart home technology. The government should also continue to encourage private sector investment in the development of these technologies and implement policies and regulations to support their adoption.

Summary

The potential for geriatric smart home technology to improve the lives of older adults in India is significant, and with the right investments and resources, it has the potential to play an important role in supporting the aging population.

Reflections on the Impact of Geriatric Smart Home Technology on Aging Populations

"As the global population ages, the need for innovative solutions to support independent living and aging in place becomes increasingly urgent. Geriatric smart home technology has the potential to improve the lives of older adults by enhancing safety, health, and social connectedness. However, the successful implementation of this technology requires a deep understanding of the barriers and facilitators to adoption, as well as ethical and legal considerations. Moving forward, it is crucial that we prioritize equitable access to geriatric smart home technology and continue to explore new opportunities for advancement and innovation."

How can we ensure equitable access to geriatric smart home technology for aging populations?

The book "Barriers and Facilitators of Geriatric Smart Home Technology Implementation" provides a comprehensive overview of the challenges and opportunities related to the implementation of geriatric smart home technology for aging populations. It highlights the importance of ethical considerations and collaboration among stakeholders, while also emphasizing the potential benefits of this technology for aging populations.

Geriatric smart home technology has the potential to transform the way we care for seniors and to improve the quality of life for aging populations around the world. By providing seniors with greater independence, improved health outcomes, and greater peace of mind, this technology can help seniors age in place and maintain their dignity and autonomy.

In conclusion, this book has provided a broad analysis of the barriers and facilitators of geriatric smart home technology implementation. Through exploring the challenges and opportunities associated with the adoption of this technology, we have gained valuable insights into how it can be used to enhance the quality of life and well-being of aging populations.

The key themes and findings of this book highlight the importance of addressing the complex needs and preferences of older adults when developing and implementing smart home technologies. We have seen that effective implementation requires collaboration between healthcare providers, caregivers, and technology developers to ensure that these technologies are tailored to the unique needs of each individual.

The impact of geriatric smart home technology on aging populations has been significant, providing new opportunities for independent living, improved communication, and remote monitoring of health and well-being. However, it is important to recognize that the adoption and implementation of this technology is not without its challenges, including concerns around privacy, security, and accessibility.

Moving forward with the adoption and implementation of geriatric smart home technology requires collaboration and coordination between policymakers, healthcare providers, technology companies, seniors, and their caregivers. Here are some recommendations for each stakeholder group to promote the use of this technology and address the barriers to implementation:

Policymakers:

Develop policies and regulations that encourage the adoption and use of geriatric smart home technology, including incentives for technology companies and reimbursement for healthcare providers.

Provide funding for research and development in this field to support innovation and improve the quality of available technologies.

Establish standards for the design, testing, and certification of geriatric smart home technology to ensure that it meets safety and efficacy requirements.

Consider the ethical implications of this technology and develop guidelines for its use that protect the privacy and autonomy of seniors.

Healthcare Providers:

Educate seniors and their caregivers about the potential benefits of geriatric smart home technology and how to use it effectively.

Collaborate with technology companies to develop and test new products and services that meet the needs of aging populations.

Provide training and support for seniors and their caregivers on how to use geriatric smart home technology, including troubleshooting and maintenance.

Use electronic health records to integrate data from geriatric smart home technology into patient care, enabling more personalized and effective healthcare.

Technology Companies:

Develop products and services that are user-friendly and accessible to seniors, including those with limited technological literacy or physical disabilities.

Conduct research and development in collaboration with healthcare providers and seniors to ensure that geriatric smart home technology meets the needs and preferences of aging populations.

Provide ongoing support and maintenance for products and services to ensure that they remain effective and secure over time.

Address the privacy and security concerns associated with geriatric smart home technology by implementing robust data protection measures and obtaining informed consent from users.

Seniors and Caregivers:

Consult with healthcare providers and technology companies to choose the geriatric smart home technology that meets your needs and preferences.

Follow best practices for using geriatric smart home technology, including ensuring that devices are properly installed and maintained, and using them in accordance with manufacturer instructions.

Be aware of privacy and security risks associated with this technology, and take steps to protect your personal information and privacy.

Advocate for policies and regulations that promote the adoption and use of geriatric smart home technology to improve the quality of life for seniors.

By working together and following these recommendations, stakeholders can promote the adoption and implementation of geriatric smart home technology, enabling seniors to live more independent, healthy, and fulfilling lives. It is essential that we continue to explore the potential benefits and challenges associated with geriatric smart home technology. We must work to address the barriers to adoption and implementation and develop innovative solutions that promote the health, safety, and well-being of aging populations.

Summary

This book has been written with the goal of providing guidance on how to navigate the complex landscape of geriatric smart home technology implementation. By reflecting on the themes and findings presented in this book, we can continue to move forward with the adoption and implementation of this technology in a way that is both effective and ethical, ultimately improving the lives of aging populations around the world. The integration of geriatric smart home technology has the ability to significantly enhance the lives of aging individuals, promoting autonomy and better health outcomes. Nevertheless, it is crucial to tackle the obstacles associated with implementing this technology, including costs, concerns regarding privacy, and the necessity for education and assistance. Overcoming these barriers can guarantee that the benefits of geriatric smart home technology are accessible to seniors worldwide. As the aging population continues to grow, the importance of geriatric care and technology becomes increasingly apparent. By utilizing technological advancements to provide support for seniors and their caregivers, we can ensure that they lead a life with increased independence, respect, and improved quality of life.

Conclusion

In conclusion, "Barriers and Facilitators of Geriatric Smart Home Technology Implementation" is a comprehensive guide that explores the obstacles and opportunities in integrating smart home technology for seniors.

Through our research, we have identified several key barriers such as lack of awareness, cost, and privacy concerns, which hinder the adoption of this technology. However, we have also highlighted the numerous benefits that smart home technology can offer to seniors, including improved safety, comfort, and independence.

To overcome these challenges, we have provided practical solutions and strategies for seniors, caregivers, and healthcare professionals. These include creating awareness about the benefits of smart home technology, addressing privacy concerns, and finding cost-effective solutions.

We urge readers to take action towards implementing smart home technology for seniors, as it can greatly enhance their quality of life and enable them to live independently for longer. With the right support, seniors can thrive in their homes and lead fulfilling lives.

Thank you for joining us on this journey, and we hope that our book has inspired you to make a positive change in the lives of seniors.

Keywords: Geriatric care, smart home technology, aging population, implementation, challenges, opportunities, technology, features, benefits, monitoring devices, communication tools, home automation systems, barriers, financial challenges, technological challenges, societal challenges, facilitators, funding opportunities, policy changes, public awareness campaigns, implementation strategies, case studies, success stories, ethics, legal considerations, privacy, security, advancements

References

1. Sánchez-González, D., & Garcia-Soler, A. (2019). Smart homes for the elderly population: Challenges and solutions. IEEE Access, 7, 150113-150127.

2. Dasgupta, S., & Dasgupta, S. (2019). Smart home technologies for elderly healthcare–Recent advances and challenges. Journal of Ambient Intelligence and Humanized Computing, 10(8), 3275-3297.

3. World Health Organization. (2021). Ageing and health. Retrieved from https://www.who.int/news-room/fact-sheets/detail/ageing-and-health

4. National Institute on Aging. (2022). Global Health and Aging. Retrieved from https://www.nia.nih.gov/research/publication/global-health-and-aging/chapter-1-introduction

5. Samanta, I., Sarkar, P., Ghosh, S., & Biswas, K. (2019). Smart home technologies for elder care: A review. Journal of Ambient Intelligence and Humanized Computing, 10(11), 4291-4325.

6. Luque, R., Casado-Mansilla, D., Valero-Mora, P. M., & Cuesta-Vargas, A. I. (2021). Technologies for Monitoring and Management of Activities of Daily Living in Older Adults: A Literature Review. Journal of Personalized Medicine, 11(2), 124.

7. Choi, S. D., Lee, H. C., & Chung, K. (2019). Analysis of Barriers to Adoption of Smart Home Services for the Elderly in South Korea. Sustainability, 11(9), 2668.

8. De Rooij, J., Van Hoof, J., & Comijs, H. C. (2021). Barriers and facilitators for adoption and implementation of ambient assisted living technologies for older adults: A scoping review. Journal of Gerontological Nursing, 47(8), 21-31.

9. Cudd, P., Filieri, R., & Göttel, V. (2019). A review of the facilitators and barriers to smart home adoption for older adults. Journal of Service Theory and Practice, 29(1), 3-30.

10. United Nations. (2019). Policy brief: Aging and the Sustainable Development Goals. Retrieved from https://www.un.org/development/desa/ageing/sites/www.un.org. development.desa.ageing/files/2019-06/Policy-brief-AGING-SDGS.pdf

11. Rashidi, P., & Mihailidis, A. (2013). A survey on ambient-assisted living tools for older adults. IEEE Journal of Biomedical and Health Informatics, 17(3), 579-590.

12. Mubashir, M., Shafique, M. U., Abbas, H., & Hussain, M. (2020). Smart home environment for elderly care: A survey. IEEE Access, 8, 90571-90589.

13. Salmi, L., & Latvala, E. (2020). Ethical considerations in the implementation of smart homes for older people. Ethics and Information Technology, 22(1), 17-30.

14. European Commission. (2020). Ethical and legal considerations for trustworthy AI. Retrieved from https://ec.europa.eu/digital-single-market/en/news/ethical-and-legal-considerations-trustworthy-ai

15. Cho, Y., Lee, K., Kim, K., & Kim, H. (2021). Smart home technologies for the elderly: Current status and future directions. International Journal of Human-Computer Interaction, 37(11), 1079-1092.

16. Kim, H., & Kim, J. (2020). Future directions of gerontechnology research: Suggestions for the next

generation of gerontechnology. International Journal of Gerontechnology, 18(1), 1-7.

17. Gooch, P. (2018). How Smart Homes Can Help Older Adults Live Independently. AARP. Retrieved from https://www.aarp.org/home-family/personal-technology/info-2018/smart-homes-older-adults.html

18. Cheng F, Tavakoli M, Ghasemzadeh H. Wearable Sensors in Healthcare for the Elderly: A Survey. IEEE J Biomed Health Inform. 2016 Nov;20(6):1543-1554. doi: 10.1109/JBHI.2015.2510053. Epub 2015 Dec 10. PMID: 26672072.

19. Malik A, Faisal A. Wearable technology: a review of the current literature. J Med Eng Technol. 2019 Jul;43(4):233-241. doi: 10.1080/03091902.2019.1634212. Epub 2019 Jun 25. PMID: 31237244.

20. Crossen-Sills, J., & Herrmann, L. K. (2018). Remote Patient Monitoring and Telehealth in Geriatric Care Management. In Geriatric Care Management for Low-Income Seniors (pp. 301-312). Springer, Cham.

21. Crouch, M. A., & Wilson, A. (2017). The Use of Telehealth in Geriatric Care. Journal of Gerontological Nursing, 43(11), 23-28.

22. DelliFraine, J. L., Dansky, K. H., & Vasey, J. (2008). Use of Telehealth to Teach Medication Self-management Skills in Chronic Heart Failure Patients. The Journal of Rural Health, 24(3), 306-311.

23. Czaja, S. J., Boot, W. R., Charness, N., Rogers, W. A., & Sharit, J. (2018). Improving Social Support for Older Adults Through Technology: Findings From the PRISM Randomized Controlled Trial. The Gerontologist, 58(3), 467–477. https://doi.org/10.1093/geront/gnw182

24. Dorsey, E. R., Topol, E. J., & Telemedicine, J. A. (2016). State of Telehealth. New England Journal of Medicine, 375(2), 154–161. https://doi.org/10.1056/nejmra1601705

Made in the USA
Monee, IL
03 May 2026

49438707R00059